The Metaphysics and Mathematics of Arbitrary Objects

Building on the seminal work of Kit Fine in the 1980s, Leon Horsten here develops a new theory of arbitrary entities. He connects this theory to issues and debates in metaphysics, logic, and contemporary philosophy of mathematics, investigating the relation between specific and arbitrary objects and between specific and arbitrary systems of objects. His book shows how this innovative theory is highly applicable to problems in the philosophy of arithmetic, and explores in particular how arbitrary objects can engage with the nineteenth-century concept of variable mathematical quantities, how they are relevant for debates around mathematical structuralism, and how they can help our understanding of the concept of random variables in statistics. This worked-through theory will open up new avenues within philosophy of mathematics, bringing in the work of other philosophers, such as Saul Kripke, and providing new insights into the development of the foundations of mathematics from the eighteenth century to the present day.

LEON HORSTEN is Professor of Philosophy at the University of Bristol. His publications include *The Tarskian Turn: Deflationism and Axiomatic Truth* (2011) and *Gödel's Disjunction: The Scope and Limits of Mathematical Knowledge* (co-edited with Philip Welch, 2016).

The Metaphysics and Mathematics
of Arbitrary Objects

LEON HORSTEN

University of Bristol

CAMBRIDGE
UNIVERSITY PRESS

CAMBRIDGE
UNIVERSITY PRESS

University Printing House, Cambridge CB2 8BS, United Kingdom

One Liberty Plaza, 20th Floor, New York, NY 10006, USA

477 Williamstown Road, Port Melbourne, VIC 3207, Australia

314-321, 3rd Floor, Plot 3, Splendor Forum, Jasola District Centre, New Delhi - 110025, India

103 Penang Road, #05-06/07, Visioncrest Commercial, Singapore 238467

Cambridge University Press is part of the University of Cambridge.

It furthers the University's mission by disseminating knowledge in the pursuit of education, learning and research at the highest international levels of excellence.

www.cambridge.org
Information on this title: www.cambridge.org/9781108706599
DOI: 10.1017/9781139600293

First published 2019
First paperback edition 2021

A catalogue record for this publication is available from the British Library

ISBN 978-1-107-03941-4 Hardback
ISBN 978-1-108-70659-9 Paperback

For CAA

Contents

Figures

Preface

For a long time I had a plan to write a book on mathematical structuralism. In this book I would put my cards on the table and try to make a contribution to the ongoing philosophical debate. But I just did not seem to be able to make solid progress with this project.

When I was honest with myself about it – and this was not very often the case – I had to admit that I did not share some of the essential presuppositions in the current debate on mathematical structuralism. At the same time, the task of re-thinking the framework of the debate was too big for me: I lacked the courage. So the book project did not really get anywhere for years.

At some point – I do not clearly remember when – some of the work of Kit Fine came to have a deep impact on me in two different ways.

First, I discovered Fine's work on the theory of arbitrary objects. I am one of the proud handful of owners of a copy of *Reasoning with Arbitrary Objects* (first edition, 1985). I was impressed by it, and instinctively felt that it had unrealised philosophical potential. Many metaphysical questions about arbitrary objects immediately came to mind, and Fine himself listed many possible applications. At the same time, I was astonished to see that Fine's theory had received so little attention in the philosophical literature.

I came to suspect that Fine's theory of arbitrary objects can be fruitfully connected to the discussion on mathematical structuralism and subsequently found out that Fine had, in an all too brief final section of his article on *Cantorian Abstraction*, already noticed this. When I read this succinct section carefully, I saw that Fine's way of relating mathematical structuralism to arbitrary object theory was not quite the way in which I would do it. In any case, as Fine himself said in his paper, the connection needed to be worked out in detail. All this gave me courage to spell out my ideas on this subject. In the course of doing this, important differences of view between Fine and me about the nature of arbitrary objects came more clearly into focus. What started as an idea for a paper grew into something that is too extensive to be crammed into a few articles.

Secondly, and just as importantly, by reading Fine's work on arbitrary object theory, I felt – there is no other way to put it – released from presuppositions in current analytical metaphysics that had bothered me for a long time. I sensed that Fine used a very different methodology, one that immediately and viscerally appealed to me. Then I found out that Fine had started to reflect on his metaphysical methodology in some of his writings from the early 2000s onward. These reflections reached (I think) a mature expression in his 2017 article on naive metaphysics. But here, too, I found that I might have something to add to what Fine wrote. I felt that Fine's departure from the received methodology in metaphysics was not radical enough: too much of the received metaphysical methodology remained in Fine's new proposal. This convinced me that I should start what had by now developed into the idea for a monograph with a chapter on the methodology of the metaphysics of mathematics. From then onwards, I felt able to make real progress in writing this book.

Fine took a shot in the dark when he wrote his book on arbitrary object theory. Today, almost 35 years later, there still is almost no philosophical literature on the subject. So I am worried that it may still be too early for a research monograph on these matters. All I can say in response to this concern is that this book is intended as an *essay* in the original sense of the word. It may well be miles off the mark: time will tell. If this book encourages more philosophers to work on the theory of arbitrary objects, then it has achieved its aim. Anyway, already now I sense an increased interest in arbitrary object theory. I take it as an encouraging sign that by the time this monograph is published, a new edition of Fine's *Reasoning about Arbitrary Objects* will (probably) finally be out.

Now that I am at it, let me relate an early influence on the writing of this book. As a visiting postdoctoral researcher, I attended a graduate seminar on modal logic by Saul Kripke in Princeton in 1999. Even though I did not know it at the time, this was a defining moment in my philosophical life. It was a strange experience. I understood very little. I was much too shy to ask questions, and I was intimidated by the environment. Foolishly, I thought that I should not spend too much time trying to understand what Kripke was trying to do but should instead concentrate on making progress with the projects that I was working on. Today I thoroughly regret not dropping all these projects completely and spending all my time on at least trying to understand what the problems were that Kripke was trying to make progress on. But Kripke's lectures influenced me more deeply than I could imagine possible at the time. As far as the present monograph is concerned, what is important is that Kripke drew attention to the expressive powers and possible applications of the Carnapian framework of quantified modal logic. (I recall that Kripke was trying somehow to prove the Fundamental

Theorem of Calculus in Carnapian quantified modal logic.) This stayed with me. Thinking about arbitrary objects from within the framework of Carnapian quantified modal logic has influenced my views on the nature and properties of arbitrary objects, as you will see.

Moving on to more recent influences, a few conversations with my former colleague Øystein Linnebo stand out in my mind. Especially his exhortation to think the philosophy of arbitrary objects through carefully, rather than to jump immediately to technical-looking applications, was important. At a conference on mathematical structuralism in Prague in 2016 he told me: *do not try to fly before you can walk.*[1] I have tried to keep that in mind. In particular, I have tried to keep my eye on the metaphysics at all times. Indeed, even though only a few applications of arbitrary object theory have been worked out in any degree of detail, what has been holding the subject back is the fact that the metaphysics of arbitrary objects is at present still ill understood and underdeveloped. We have jumped too quickly to applications.

There is not much background literature to fall back on, so most of the material in this book is new. Nonetheless, there are a few places in this book where I use material from articles that I have published or are in press. In Chapter 8, I draw in places on my article Horsten (2019). Chapter 9 is partly based on my article Horsten and Speranski (2018). In Chapter 10 I rely on a theory of non-Archimedean probability that I have developed together with Vieri Benci and Sylvia Wenmackers (Benci et al. 2013) and on an unpublished article in which this framework is applied to the set theoretic universe (Brickhill and Horsten 2019).

---

[1] My only complaint to Øystein is that he used the word 'walk' instead of 'crawl', or 'roll over'.

Acknowledgements

I am indebted (in no particular order) to Kit Fine, Hazel Brickhill, Stewart Shapiro, Philip Welch, Stanislav Speranski, Øystein Linnebo, Ofra Magidor, Ryo Ito, Richard Pettigrew, Sakaé Fuchino, Thomas Müller, Jan Heylen, Rafael De Clercq, Johannes Stern, Catrin Campbell-Moore, Gideon Rosen, John Burgess, Volker Halbach, Timothy Williamson, Oliver Tatton-Brown, and Giorgio Venturi, who have generously commented, in discussion or via email, on my ideas when they were in an early stage of development. I hope I have not left anyone out.

I have presented material in this book at various conferences and workshops, and I am grateful for all the comments that I have received. In particular, I benefited from encouraging comments and suggestions by members of the audience at the conference on *The Emergence of Structuralism and Formalism* (Prague, 2016), where I first presented some of my ideas on the connection between arbitrary object theory and mathematical structures. Special thanks to Oliver Tatton-Brown and Volker Halbach for carefully going through a version of the entire manuscript and generously commenting on it.

This book was written mostly during the academic year 2017–2018, when I was on research leave from the University of Bristol. I spent the first term (Michaelmass, 2017) as a Visiting Fellow in Oxford. Attending Tim Williamson's postgraduate metaphysics seminar during that period played an important role in shaping my views on metaphysical method. I will never forget how I ended up, completely by accident, with Tim Williamson, Ofra Magidor, David Wiggins, and Volker Halbach at a small coffee table in the Senior Common Room in New College. As Volker remarked: where else in the world could such a thing happen? I spent the rest of that academic year as a Visiting Professor in Kyoto University. In this wonderful environment most of this book was written. I am immensely grateful to Professor Tetsuji Iseda for the welcoming academic environment that he provided, and for his kind invitation to give a research seminar lecture (CAPE lecture) on metaphysical methodology. I also enjoyed all sessions of the logic and metaphysics reading group (organised by Ryo Ito and Takuro Onishi) in Kyoto. And I am very grateful to the logicians at Kobe University (especially to Professor Sakaé Fuchino) for their interest in my work during my time in Japan.

In July and August of 2017, Stanislav Speranski was in Bristol on a Benjamin Meaker Visiting Professorship. I had an excellent time collaborating with him on the formalisation of reasoning about the generic ω-sequence in quantified Carnapian modal logic.

I thank the members of the Foundational Studies Bristol group (especially Catrin Campbell-Moore and Johannes Stern) for their comments at a research seminar that I gave to that group in 2017 on generic systems.

Hilary Gaskin of Cambridge University Press has been very patient with me during the long time when I was not able to make real progress with the manuscript of this book. I also want to express my gratitude to Sophie Taylor of Cambridge University Press. Together with Hilary Gaskin, she provided invaluable assistance and encouragement in the last stages of the writing of this book.

Addressing you, Hazel, is the most difficult: I do not know where to begin. Putting up with my philosophical questions to you for a couple of years is the least of it. You got me out of philosophical confusions on so many occasions, and showed me where and how conceptional distinctions needed to be made. There are just so many things that we have thought through together by talking about it to each other. (I don't know what you think of it all.)

Symbols and Abbreviations

1 | Introduction

1.1 Arbitrary Objects and the Philosophy of Mathematics

Mathematics is the science of structure. Such is the battle cry of structuralism in the philosophy of mathematics.

Mathematical structuralism emerged on the scene as a major position in the philosophy of mathematics in the writings of Dedekind and Hilbert at the end of the nineteenth century. Since then, the discussion of mathematical structuralism has renewed itself several times. Around 1940, Bourbaki connected mathematical structuralism with set theory. In 1965, Benacerraf articulated an enormously influential argument for mathematical structuralism. In the 1980s, new forms of mathematical structuralism were developed: nominalistic ones (Field, Hellman) and platonistic ones (Resnik, Shapiro).

Structuralism became the dominant view in the philosophy of mathematics during the 1980s. Since then, and as a result of this development, the philosophical discussion about mathematical structures has become more inward looking: it has become to a very considerable extent an *internal* debate between different forms of mathematical structuralism. In recent years, despite a growing influence of ideas from category theory on the philosophical debate, the discussion about mathematical structuralism seems to be running out of steam. New ideas are needed if we are to rejuvenate it once more.

The discussion among mathematical structuralists has come to revolve to a large extent around a question of realism. Should we conceive of mathematical structures as platonic entities, or are mathematical structures reducible to systems of objects that can be recognised from a non-platonistic standpoint to exist?

This concentration on questions of mathematical realism has taken our eye off the question of the *nature* of mathematical structures. It has come to stand in the way of developing deeper insight in the metaphysical nature of mathematical structures. This book is an attempt to move away from the question of realism and to strike out in a new direction. One of my aims is to explore a new way of understanding mathematical structures.

Well, the idea of how to do this is not completely new. It is inspired by a metaphysical theory that takes the idea of *arbitrary objects* at face value. The thought is that there are specific objects, such as you and I, but there are also arbitrary objects such as *the man on the Clapham omnibus*. This view goes back at least to the nineteenth century, but it was developed into a metaphysical theory by Kit Fine in the 1980s. In the philosophy of mathematics he mainly applied it in his theory of Cantorian abstraction, but in a last brief section of an important article he even applied it to mathematical structuralism.

The idea of bringing arbitrary object theory to bear on the question of the nature of mathematical structures is, very roughly, this: mathematical structures stand to systems of objects that instantiate them as arbitrary objects stand to the specific objects that can be their values. Just as we make a metaphysical distinction between specific and arbitrary objects, we distinguish between specific systems and *generic systems*, which is what structures are.

Fine's theory of arbitrary objects did attract attention in the natural language semantics community, but less so in the metaphysical community, although now it is gradually becoming clear how much philosophical potential his theory has. Fine's suggestions for applying this theory to mathematical structuralism have attracted even less attention by the philosophical community. Given that Fine is widely recognised as one of the leading metaphysicians of our days, this is surprising.

In fairness to the philosophical community, I think the scant attention given to Fine's theory of arbitrary objects is due in part to the fact that Fine has not worked out his theory of arbitrary objects in sufficient metaphysical detail. In his (overall very positive) review of Fine's book *Reasoning with arbitrary objects*, Santambrogio (1988, p. 634) writes:

The main criticism I have is that a richer discussion at an informal level of the philosophical status of arbitrary objects could have been more interesting than some of the technical results he has produced.

Santambrogio is right. Logical questions about reasoning with arbitrary objects are important, and so are philosophical applications of arbitrary object theory. But the metaphysics of arbitrary objects is difficult and subtle, and Fine's discussion of it is dense at places. Readers have expressed difficulties in understanding aspects of Fine's theory of arbitrary objects.

Most philosophers today still find the arguments of Frege against arbitrary objects convincing, and they find the proposal of taking the idea of arbitrary objects at face value a bit mad. But, equally, analytic philosophers generally do not object to a philosophical view being crazy – especially if it is a metaphysical view – as long as they feel that they understand it. Therefore I think that the metaphysics of arbitrary objects should be given a more extensive, in-depth treatment, and deserves a more perspicuous and systematic development than one finds in Fine's writings. Taking this to heart, I will devote much attention to the embedding of my discussion of generic systems of mathematical objects in the general framework of the theory of arbitrary objects.

In any case, I believe that Fine's ideas in this area deserve a better fate and want to play a role in bringing this about. I aim to develop Fine's ideas into a more fully developed metaphysical theory of mathematical structure. My view of the nature of mathematical structures is not the same as that of Fine. This is to a significant extent because Fine and I disagree about the way in which the underlying theory of arbitrary objects should be de developed. Nonetheless, you can find most of the basic ideas that are developed in this book already in Fine's work.

1.2 Scratching the Surface

Arbitrary object theory is a very young sub-field of analytical metaphysics. The philosophical and logical literature on the subject is very limited: philosophical unclarity abounds, and there are myriad unsolved logical questions. This makes writing a monograph on arbitrary object theory a daunting enterprise. There are moments when one has a sense of being overwhelmed – although some would say that this is simply an illusion: there is nothing there to be seen at all.

My objectives in this monograph must therefore necessarily be modest. They are threefold. In the first place, I want to develop a metaphysical perspective on the nature of arbitrary entities and to defend it against objections. Secondly, I want to develop a few applications of arbitrary object theory to some extent. Thirdly, I want to develop a perspective on how various themes in the contemporary literature on arbitrary objects are connected with each other.

In his book *Reasoning about Arbitrary Objects*, Fine has provided robust philosophical replies to the strongest objections against arbitrary objects (most of which were first articulated by Frege). Moreover, Fine has also worked out elements of a metaphysical theory of arbitrary objects. But the metaphysical theory remains underdeveloped: the nature of arbitrary objects is presently not well understood. Articulating the metaphysical picture of arbitrary objects in some detail is therefore my main objective.

In broad outlines my theory agrees with that of Fine. But I want to develop it further, and there are important aspects in which my account differs with Fine's.

Turning to the second objective, I will be mainly interested in the way in which arbitrary object theory can be applied not only to objects but also to *systems* of objects. I will be concerned with *arbitrary* systems of objects, which, it turns out, can also be seen as systems of arbitrary objects. Moreover, I will also discuss the extent to which arbitrary objects can be said to be more or less *likely* to have some property. As far as attributing probabilities to arbitrary objects is concerned, we will see that there are problems relating to arbitrary objects that can take on infinitely many values that I am not able to solve in a completely satisfactory manner.

Concerning the third objective, I aim to show how some of the main strands in the small body of literature about arbitrary objects from the past four decades are related to each other. Again, I cannot claim to have completely succeeded. In particular, I regret not having been able fully to integrate the interesting theory that was developed by Santambrogio (1987) in this book. Nonetheless, I believe that the bibliography at the end of this book is at present fairly complete concerning philosophical thought on arbitrary objects from around 1983 until 2018 – and hope that this bibliography will soon be labelled as dated.

On the whole, I am only able to scratch the surface. I seek an understanding of what will be my stock example: the generic system of the natural numbers. I will extrapolate from this example to other examples and to the shape that a general theory of arbitrary objects could take. But I will not be able to go far beyond an understanding of the arbitrary natural numbers and the structures to which they belong. So I will fall far short of developing a general metaphysical and logical theory of arbitrary objects and generic systems.

In sum, the ambition of this book is to lay out some of the groundwork for a theory of arbitrary objects. It is a *prolegomenon to a future theory*. Even if my account does not go too far off track, there is an immense amount of work that remains to be done. And there is indeed a great risk that I have gone off track at crucial junctures. Most mistakes in arbitrary object theory have not yet been made, so I am very likely to make some of them.

1.3 Structure and Method

This book is divided into 11 chapters, but at a higher level it can be roughly be subdivided into three parts. The first part discusses arbitrary object theory as an exercise in what Fine calls *naive metaphysics* (Chapters 2–4).

The second part investigates the relation between the notion of mathematical structure and arbitrary object theory (Chapters 5–9). The third part is shorter: it explores the connection between arbitrary objects and random variables, and the concept of probability that goes with it (Chapter 10). (The brief Chapter 11 contains a discussion of open problems and avenues for future research.) Throughout the book, the natural numbers and the structure to which they belong are taken to constitute the paradigmatic test case for the problems and theories that are discussed.

Let me now turn to a more detailed overview of the structure and methodology of the book.

The fact that questions of realism have come to play a central role in the philosophy of mathematics, as they have done in other philosophical disciplines, such as philosophy of science and philosophy of mind, is no accident. It is the result of a methodology that has become prevalent in analytic philosophy, largely through the work and influence of Quine. I start in the next chapter (Chapter 2) by advocating an alternative methodology in metaphysics. Inspiration for this effort can be found in Fine's writings. I rely on his distinction between naive and foundational metaphysics. Naive metaphysics is concerned with the way things metaphysically appear to us, whereas foundational metaphysics is concerned with which entities exist. In a word, Fine's slogan is:

> *Naive metaphysics first.*
> *Foundational metaphysics second.*

I go further than this and argue that we should not merely postpone questions of existence until after inquiring into their metaphysical nature; questions of existence should be dismissed altogether. In other words, I argue that a version of the forms of quietism that have been proposed in realism debates in other areas of philosophy (such as philosophy of science and philosophy of mathematics) can be and should be adopted *in* metaphysics.

As a naive metaphysician, I am sceptical about most ontological reductionist programmes in philosophy. In particular, I strongly reject the thesis that arbitrary entities can somehow be reduced ontologically to classes of specific objects. This does not mean, however, that I am sceptical about the use of mathematical methods in metaphysics. Quite on the contrary: I see no contradiction between naive metaphysics and 'mathematical philosophy'. Indeed, the success of the whole enterprise will to a significant extent hinge on how well the set theoretical models that I will use capture salient logical properties of and relations between arbitrary entities. Understanding the nature of arbitrary entities and being able to model them well go hand in hand.

I want to practice what I preach by *doing* naive metaphysics. In particular, Chapter 3 contains an investigation into the *nature* of arbitrary objects. The account that I propose differs in key points from that of Fine, but a detailed comparison with his account is postponed until Chapter 7. Instead, in Chapter 3, we go further back in history. I show how key elements of the theory of arbitrary objects (as I develop it) are already present in the theory of *variables* that was developed by Russell in his *Principles of Mathematics* (1903).

The paradigmatic examples of arbitrary objects have always come mostly from mathematics. Chapter 4 contains an application of arbitrary object theory to metaphysical questions about mathematical objects. Special attention is given in this chapter, as in the remainder of the book, to the natural numbers. Moreover, I investigate to what extent arbitrary object theory is connected with the theory of what Charles Parsons calls *quasi-concrete* objects such as linguistic expressions.

In Chapter 5 I turn to philosophical questions about mathematical structures. If my arguments in Chapter 2 are sound, then the realism debate about mathematical structures is misguided. Nonetheless, my aim cannot be achieved without a thorough discussion of an ongoing realism debate, viz. the dispute between platonistic and nominalist versions of mathematical structuralism. These accounts contain essential insights into aspects of the nature of mathematical structures. They are a source of adequacy conditions that any good theory of mathematical structure will have to satisfy. In particular, a good theory of mathematical structure should leave room for a concomitant theory of mathematical objects, it should not be vulnerable to a variant of Benacerraf's identification problem, and it should admit a plausible account of what it means to compute on the natural numbers. Later in the book, my account of mathematical structures is tested against these conditions.

One central claim of this book is that in the same way as there are, beside specific objects, also arbitrary objects, there are also, beside specific systems of objects, arbitrary systems of objects. Arbitrary objects stand to specific objects in roughly the same way as arbitrary systems of objects stand to specific systems of objects:

$$\frac{\text{arbitrary object}}{\text{specific object}} \approx \frac{\text{arbitrary system}}{\text{specific system}}.$$

Moreover, there is a special subclass of the class of arbitrary systems, which I will call *generic* systems. Another central thesis of this book is then that generic systems can be identified with mathematical structures and vice versa. In this way, in Chapter 6 I connect the theory of arbitrary entities with the question of the nature of mathematical structures.

I will propose an account according to which many mathematical theories – algebraic theories as well as non-algebraic theories – can be said to describe one or more generic systems. Foundational mathematical theories, however, turn out to be exceptions. If a discipline such as set theory can be said to describe arbitrary entities at all, then it must be in a sense that is different from the way in which non-foundational mathematical theories can be taken to be about arbitrary entities.

Chapter 7 compares Kit Fine's metaphysical theory of arbitrary objects and generic systems with the account that I propose in earlier chapters of the book. It will emerge that the most important difference between Fine's account and mine concerns the relation of *dependence* between arbitrary objects. On Fine's account, dependence is a directional and irreducible cornerstone notion of arbitrary object theory. In my account dependence between arbitrary objects is a definable relation.

Chapter 8 is concerned with the relation between my conception of mathematical structures as generic systems and existing forms of mathematical structuralism. The natural number structure functions as a paradigmatic test case. I show how generic systems such as the natural number structure can be taken to be composed of mathematical objects. I argue that my account is invulnerable to Benacerrafian identification problems and that it can make philosophical sense of the fact that we compute *on* the natural numbers.

However, it will emerge that my account should not be seen as a rival to existing forms of mathematical structuralism. I do not take a stance in the realism debate about mathematical structures: '-isms' in general have too much of a reductive flavour for my taste. Moreover, there are good reasons for resisting the rather too broad claim that the subject matter of mathematics *is* mathematical structures.

The remaining two chapters in the book are somewhat more technical. They can be seen as mainly concerned with following up two logical suggestions that were made by Saul Kripke.

In Chapter 9 I carry out a logical investigation of one particular generic system: the structure of the natural numbers. This turns out to be a very rich mathematical structure. Following a suggestion of Kripke, it is argued that *Carnapian quantified modal logic* is the appropriate formal framework for expressing facts about specific and arbitrary numbers, for reasoning about them, and for formulating and investigating key concepts pertaining to arbitrary objects (such as indistinguishability and dependence). The main result of this chapter is that arbitrary numbers can play the role of *sets* of natural numbers, i.e. of real numbers. There is a precise sense in which the theory of the natural number structure, seen as a generic system, is exactly as expressive as the framework of second-order number theory.

Kripke once suggested that *random variables* can be seen as Carnapian individual concepts. Carnapian individual concepts will be seen to be closely related to, but not quite the same as, arbitrary objects. In Chapter 10, the analogy between random variables and arbitrary objects is probed. The leading thought is that the theory of random variables suggests a natural way in which probabilities can be associated with arbitrary objects. For arbitrary objects with a finite associated value range of specific objects, it is not hard to see how one should proceed. For arbitrary objects with infinite value ranges, however, it is less clear what a satisfactory notion of associated probabilities would look like. In this context, I will turn to *non-Archimedean probability theory* as the best account that I can come up with.

In a short final chapter, I try to step back and reflect on the *scope* of the arbitrary object theory that I have developed. Moreover, some possible avenues for further research are highlighted and briefly discussed. The list of problems that I discuss is far from exhaustive.

By the time that the reader has arrived at the end of the book, she will be keenly aware of the fact the metaphysics and logic of arbitrary objects still remain in an underdeveloped state, and that many of the issues that have been touched upon in this book have not been given the attention that they ultimately deserve. I can only express the hope that this will motivate the reader to try to develop the theory further and to correct some of my mistakes.

1.4 Intended Audience

This is a research monograph rather than a textbook or a reference text. It is aimed at graduate (or post-graduate, in UK terminology) students in metaphysics, philosophy of mathematics, and/or philosophical logic and at professional philosophers with a research interest in these domains.

Not every reader may want to read the whole book. If you are only interested in the metaphysics of arbitrary objects and generic systems, then you may may simply stop reading after Chapter 8. If you only want to see the methodology that I use practiced rather than discussed, then you can skip

Chapter 2. If you are already sufficiently familiar with the intricacies of the debate between eliminative and non-eliminative structuralists in the philosophy of mathematics, then you may not want to read Chapter 5.

If you are someone who is primarily interested in the logico-mathematical part of this book, then the quickest way is to start with Chapter 6 before going on to Chapters 9 and 10. But this may leave you wondering why the logical framework that is used is the right one for the task at hand. So in this case I recommend you to skip Chapters 2 and 5, but at least have a cursory look at Chapters 3, 7, and 8 before moving to the technical part of this monograph.

I attempt to be as clear as possible and to make this monograph reasonably self-contained – I found complete self-containment unachievable. If you want to get the most out of this book, and especially if you want to adopt a healthy critical attitude towards its content, it helps if you are able to situate it in the wider context in which it is embedded.

On the philosophical side, I assume first of all familiarity with contemporary literature on the philosophy of mathematics. Especially, I assume readers to be at least somewhat familiar with the main themes in the contemporary debate on mathematical structuralism. I expect them to be acquainted with what is covered in a good introduction to the philosophy of mathematics such as Linnebo (2017). Familiarity with the ground that is covered by more advanced works such as Shapiro (1997) or Parsons (2008) – which cannot really be called textbooks – is more than sufficient. Secondly, a significant part of the action in this book takes place in and around the intersection of philosophy of mathematics and logic. An excellent work that covers all the background that is relevant here (and more), is Button and Walsh (2018). Thirdly, I also assume a good understanding of the main concepts and themes of contemporary analytical metaphysics. Familiarity with what is covered in a book such as Loux and Zimmerman (2003) definitely suffices.

On the logical side, I presuppose what is covered in basic and intermediate courses on mathematical and philosophical logic. As far as mathematical logic is concerned, I rely on knowledge of basic proof theory, model theory, and the relations between the two. Boolos et al. (2007) covers almost all of this, except that I will also make use of arguments involving ultrafilters. For an introduction to the latter, the reader is referred to Bell and Slomson (2006). As far as philosophical logic is concerned, I assume knowledge of intermediate philosophical logic; in particular knowledge of the main concepts and techniques of quantified modal logic. Knowledge of the material in Garson (2006) is more than enough.

To conclude, familiarity is assumed with basic elements of some mathematical disciplines. Only very elementary set theory is presupposed: Schoenfield (1977) more than suffices. Concerning the foundations of number theory and analysis, Truss (1997) is recommended, and some basic knowledge of non-standard analysis as is contained, for instance, in the first three chapters of Goldblatt (1998), is useful. Concerning probability theory, Blitzstein and Hwang (2015) more than suffices. In addition, familiarity is assumed with basic elements of graph theory: the first chapters of Diestel (2006) will do.

1.5 On Notation and Technical Matters

Since metaphysical and logical issues are intertwined in this book, it is written in a mixture of prose and technical notation. Throughout the book, I try to use standard logical and mathematical notation. I try to be careful in describing new notation clearly at the point where it is introduced. A glossary of symbols and abbreviations can be found at the end of the preamble to this book.

I use standard terminology for logical concepts:

- $\wedge$ stands for conjunction;
- $\vee$ stands for disjunction;
- $\neg$ stands for negation;
- $\rightarrow$ stands for material implication;
- $\leftrightarrow$ stands for material equivalence;
- $=$ stands for numerical equality;
- $\Diamond$ stands for possibility (and $\Box$ stands for necessity).

As is also usual, I use the symbols $\wedge$, $\vee$, $\overline{(\ldots)}$ for the corresponding operations of Boolean algebras. It will be clear from the context when for instance $\vee$ stands for the 'join' operation in some Boolean algebra rather than the more specific logical operation of disjunction.

The distinction between specific and arbitrary objects will be of fundamental importance. In prose contexts, I mostly use lowercase letters from the end of the Roman alphabet $(m, n, \ldots)$ to refer to *specific* objects and lowercase letters from the beginning of the Roman alphabet $(a, b, c, \ldots)$ to refer to *arbitrary* objects.

When I *model* arbitrary objects in set theory as functions – as will often happen in technical contexts – I refer to them by lowercase Greek letters

$(\alpha, \beta, \gamma, \delta, \dots)$. I use the usual symbols for elementary set theoretic operations and relations:

- $A \cap B$ stands for the intersection of the sets A and B;
- $A \cup B$ stands for the union of the sets A and B;
- $A \times B$ stands for the Cartesian product of A and B;
- $A \subseteq B$ expresses that A is a subset of B;
- $A \backslash B$ stands for the set of elements of A that do not belong to B;
- $\emptyset$ stands for the empty set.

When I use functions that are not intended to refer to arbitrary objects, I use the ordinary *math font* symbols $f, g, \dots$ The expression $dom(f)$ stands for the domain of the function f; $ran(f)$ stands for the range of f.

I use capital *sans serif* letters ($\mathsf{A}, \mathsf{B}, \dots$) to refer to *sets* of arbitrary objects. And I use capital boldface letters ($\mathbf{N}, \mathbf{G}, \dots$) to refer to arbitrary *systems* of objects, whereas I use *math font* capital letters ($S, T, \dots$) to refer to *specific* systems of objects.

Upper case Latex *math font* letters $A, B, \dots$ stand for sets. Lowercase Greek letters are used not only for arbitrary objects modeled as functions (as explained above) but also for ordinal numbers. I rely on the context to make it clear when a lowercase Greek letter is used for an ordinal and when it is used for an arbitrary object. The expression B^A stands for the collection of functions from the set A to the set B. In particular, $\omega^{<\omega}$ stands for the collection of functions from finite sets of numbers to ω. Cardinal numbers are identified with the smallest ordinals equinumerous with them. The ordinal ω, for instance, will stand for the smallest infinite cardinal number, and I will identify the cardinal of 2^ω, i.e. of the continuum, with the smallest ordinal of that cardinality. The cardinality of a set A is denoted as $|A|$. (But sometimes I will be sloppy. For instance, I will often write 2^ω for $|2^\omega|$.) The symbol On stands for the class of ordinals. The symbol V stands for the set theoretic universe, and V_α stands for the ordinal rank α thereof. The symbol $\mathcal{P}$ refers to the full power set operation, and $[S]^{<\kappa}$ refers to the subsets of the set S that are of cardinality $<\kappa$.

Concerning proof theoretic notation, I use the lowercase letters $x, y, z, \dots$ as first-order variables, and the uppercase letters $X, Y, Z, \dots$ as second-order variables. Calligraphic font $\mathcal{L}, \mathcal{L}_1, \dots$ is used to refer to formal languages, and Greek letters $\varphi, \psi, \dots$ are used to refer to formulas of a formal language. $\varphi[y\backslash x]$ stands for the uniform substitution of all free occurrences of the variable x by occurrences of the variable y in the formula φ. As usual, the symbol $\vdash$ stands for the derivability relation. I use uppercase Roman letters to refer to theories, which are intended to be closed under logical derivation. For instance, PA stands for first-order Peano arithmetic, PA^2 stands for second-order Peano arithmetic, ZFC stands for first-order Zermelo-Fraenkel set theory with the Axiom of Choice, and ZFC^2 stands for second-order Zermelo-Fraenkel set theory with the Axiom of Choice.

The *fraktur font* $\mathfrak{A}, \mathfrak{B}, \ldots$ is used to refer to models. As is common, the symbol $\models$ stands for the model theoretic 'making true' relation. The notion $\mathfrak{A} \models \varphi[x/d]$ expresses that in the model $\mathfrak{A}$, the formula φ is made true on the assignment that assigns the element d of the domain of $\mathfrak{A}$ to free occurrences of the first-order variable x.

Familiar mathematical structures will be referred to in familiar ways. For instance, $\mathbb{N}$ refers to the standard natural number structure, $\mathbb{Q}$ refers to the structure of the rational numbers, $\mathbb{R}$ refers to the structure of the real numbers, and $\mathbb{C}$ refers to the structure of the complex numbers.

State spaces will play an important role in my account of arbitrary objects and in my application of arbitrary object theory to random variables. I will mostly use the symbol Ω to refer to state spaces, and I will use variables $X, Y, \ldots$ for random variables. $\Pr(A)$ will stand for the (unconditional) probability of the event A, and $\Pr(A \mid B)$ will stand for the probability of the event A *conditional* on the event B.

Even though technical aspects cannot and will not be side-stepped, this is primarily intended to be a philosophical work. Conceptual and meta-physical questions take precedence over mathematical questions. Therefore the reader will not find many logical or mathematical theorems in this monograph, and even fewer proofs. Only simple propositions are proved, sometimes in a sketch way. For non-elementary proofs of theorems, the reader is referred to the literature.

2 | Metaphysics of Mathematics

> Art is not about something. Art is something.
>
> *(Sarah Bernhardt)*

This chapter deals with methodological questions. I will not use the methodology that is typically adopted in contemporary analytical philosophy, especially in metaphysics, and in philosophy of science and neighbouring fields. The methodological approach that is advocated instead in this chapter leaves little room for realism debates in philosophy. In particular, from the next chapter onward, questions of realism in metaphysics and in philosophy of mathematics are left mostly aside. Instead, I will be doing what Kit Fine calls *naive metaphysics*.

2.1 The Methodology of Philosophy of Mathematics

There is a typical way of proceeding, or methodology, that has very often been practiced by philosophers of mathematics since the latter part of the twentieth century. It goes roughly as follows.

As a starting point, a body of mathematical knowledge is taken as given. It takes the form of a collection of mathematical theories. The body of knowledge that the philosopher of mathematics takes as given typically includes arithmetic and analysis, and set theory. Less attention is usually given to theories such as group theory, algebra, topology, or graph theory. Mathematicians develop these theories. Working 'in' these theories, they produce mathematical knowledge in the form of definitions, proofs, and theorems. But they also talk informally about mathematical theories, the relations they bear to each other, and what they are about. This is also taken as 'data' by the philosopher of mathematics.

Quine famously argued that the philosopher should try to make philosophical sense of our best scientific theories rather than to question them or parts of them (Quine 1981, p. 21):

[I]t is within science itself, and not in some prior philosophy, that reality is to be identified and described.

Quine's stance has had an enormous influence on the methodology of metaphysics ever since the 1960s. Sider, who shares Quine's methodological viewpoint, puts it well (Sider 2011, pp. 166–167):

The most promising thought about methodology in ontology, to my mind, is Quine's ... exhortation to believe locally by thinking globally – to evaluate local ontological claims by assessing global merits of theories that contain them. These global merits include such factors as 'simplicity', as well as fit with and prediction of our evidence ... One finds this methodology, in one form or another, throughout contemporary writing on ontology.

In the spirit of Quine's *scientific naturalism*, the philosopher of mathematics regards it as her business to make philosophical sense of our best mathematical theories – which are indispensable to natural science – rather than to question them or part of them. In Shapiro's (1997, p. 110) words:

Advocates of some contemporary strategies thus propose to square mathematics with naturalized epistemology by showing that the boundary of the natural extends to include the mathematical. Numbers, points, Hilbert spaces, and maybe even the set-theoretic hierarchy are natural objects within the bounds of ordinary scientific scrutiny. Mathematics is part of science and cannot be exorcized from it.

Two philosophical questions then present themselves:

1 What are mathematical theories about?
2 How do we acquire mathematical knowledge?

The first of these questions is metaphysical; the second is epistemological. The answers to these questions are constrained by what the mathematical community actually takes to be facts belonging to, or about, established mathematics.

Concerning the second question, a straightforward and obvious answer is found in mathematical practice itself: mathematical knowledge is acquired by *mathematical proof*. A mathematical proof is an argument that (ultimately) deduces mathematical facts from basic mathematical axioms in a logically valid manner. Many believe that we have an exhaustive and precise list of the principles and rules of logically valid reasoning. They are the basic laws of *classical first-order logic*: choose your favourite deduction system. Moreover, many believe that we know *why* these principles and rules are valid: they are necessarily truth-preserving. But there is no agreement on whether we have a complete list of the basic mathematical axioms. There is not even agreement about whether having a *complete* list of mathematical axioms is even possible in principle. Moreover, even for the core of basic axioms that are accepted by the vast majority of mathematicians, there is no agreement on what our epistemological justification for our belief in these basic axioms is. Indeed, much of the debate in mathematical epistemology of the past half-century is about what our justification for what we take to be basic mathematical axioms consists in. *This* book, however, is more about mathematical *metaphysics* than about mathematical epistemology. So I will be more (but not exclusively!) concerned with question 1 than with question 2.

We may try to answer question 1 in a straightforward and deflationary manner as follows. Mathematics is about lots of different kinds of things. Mathematics is about many kinds of numbers (e.g. natural numbers, fractions, real numbers, complex numbers, *p*-adic numbers), spaces (e.g. topological spaces, vector spaces, measure spaces), algebraic structures (e.g. groups, rings), sets, mappings, algorithms, graphs, and so on.

But here ontological questions appear rather quickly. What is the nature of these objects? Are they fundamentally somehow similar in nature? For some of the elements in the list we may ask if they are really objects rather than something else. Perhaps an algorithm, for instance, is a process rather than an object. Moreover, is not mathematics in some sense also about *concepts*, such as the concept of continuity, the concept of smoothness, or the concept of computation? Does an *ontological reduction* relation hold between some of these objects? For instance, are Abelian groups perhaps *really* sets of some kind?

In fact, many philosophers of mathematics will insist that a straightforward approach to answering the question of the subject matter of mathematics is misguided. Metaphysical questions about the subject matters of mathematical theories are taken to presuppose questions about the *meaning* of statements belonging to the theories (Dummett 1993, Introduction). Barring constructivist worries, it is virtually always assumed that *classical quantificational logic* plays a role in the meaning of mathematical statements. But it is up for discussion whether the coarse surface grammar of mathematical statements as given by their 'naive' formalisation in first-order logic respects their real meaning. Perhaps *higher-order quantification* is part of the 'deep' meaning of mathematical statements. Perhaps *modal concepts* are involved in the content of mathematical statements. In sum, the meaning of mathematical statements is taken to be debatable: mathematical statements do not have to be taken at naive face value, whatever that might exactly amount to.

At this stage of the investigation, questions of *semantic reduction* can come to the fore. For instance, it may be claimed at this point that the meaning of mathematical terms can be given using only physical vocabulary and a modal notion (Hellman 1993). If that is right, then mathematics is ultimately about (possible) physical objects, properties, and relations.

Despite hopes of the logical empiricists to the contrary, when the philosopher of mathematics has arrived at her preferred account of the meaning of mathematical statements, there will be deep metaphysical questions left. If, for instance, mathematical discourse can be semantically reduced to discourse about possible physical objects, then a philosophical account of the nature of the modality involved is required. If the philosopher of mathematics has concluded that a substantial part of mathematical discourse can be reduced to talk about space-time regions (Field 1980), then a metaphysical account of the nature of space-time regions is needed. Moreover, in this case there will be questions about ontological reduction left. Might spatiotemporal properties be reducible to an ontologically more fundamental basis: to point events and relations of causality, for instance (Sklar 1975)?

Crucially, the account of the meaning of mathematical statements that is adopted must of course also be subjected to philosophical scrutiny. There will often be a *trade-off*. The more tractable the remaining metaphysical questions are, the more questionable the account of the meaning of mathematical statements tends to be, and vice versa.

Many contemporary general metaphysicians are less inclined to tolerate a large divergence between surface grammatical structure and 'deep' grammatical structure of statements of science than metaphysicians used to be. It is felt that there is a very straightforward transition from the naturalistically given 'data' to the metaphysical questions: the meaning theoretic stage of the investigation can largely be skipped. But philosophers of mathematics have, until now, been reluctant to follow this trend: most contemporary theories of the metaphysics of mathematics contain a very substantial meaning theoretic component.

2.2 Quietism

Kit Fine gives the following methodological characterisation of how metaphysical investigations are mostly conducted in contemporary analytical philosophy. Contemporary metaphysics can in his view be roughly divided into two parts (Fine 2017b, p. 98):

1 ontology
2 metaphysics proper

Ontology aims at arriving at judgements about *what there is*. Metaphysics proper investigates the *nature of what there is*. The metaphysician has to investigate what exists *before* enquiring into its nature. Ontology precedes metaphysics proper because what does not exist does not have a way of existing that can be investigated philosophically. Fine calls this metaphysical practice *traditional metaphysics*.

The typical methodological pattern in philosophy of mathematics that was sketched in Section 2.1 corresponds to this methodological pattern in metaphysics. First figure out what the mathematical world consists of, then find out what the nature of these constituents is. Figuring out what the mathematical world consists of involves taking formal and informal mathematics to be largely true.[1] But the *meaning* of commonly accepted mathematical statements is very much up for philosophical debate.

A number of influential philosophers have felt uncomfortable with this way of doing metaphysics. Let us look at a few ways in which philosophers have expressed this discomfort.

Carnap (1950, pp. 206–207) introduced a distinction between internal and external questions concerning science and mathematics in the context of his theory of linguistic frameworks:

It is above all necessary to recognize a fundamental distinction between two kinds of questions concerning the existence or reality of entities. If someone wishes to speak in his language about a new kind of entities, he has to introduce a system of new ways of speaking, subject to new rules; we shall call this procedure the construction of a linguistic framework *for the new entities in question. And now we must distinguish two kinds of questions of existence: first, questions of the existence of certain entities of the new kind within the framework; we call them* internal questions; *and second, questions concerning the existence or reality of the system of entities as a whole, called* external questions. *Internal questions and possible answers to them are formulated with the help of the new forms of expressions. The answers may be found either by purely logical methods or by empirical methods, depending upon whether the framework is a logical or a factual one. An external question is of a problematic character which is in need of closer examination ... From the*

---

[1] Mathematical fictionalism is an exception.

[internal] questions [about things] we must distinguish the external question of the reality of the thing world itself. In contrast to the former questions, this question is raised neither by the man in the street nor by scientists, but only by philosophers.

So internal questions can be answered by the methods of the theory. External questions *sound* like internal questions that have *obvious* answers, but the questioner will protest that that is not what she means. But then, in Carnap's (1950, pp. 206–207) view, they are from the point of view of the linguistic framework *meaningless*:

Therefore nobody who meant the question 'Are there numbers?' in the internal sense would either assert or even seriously consider a negative answer. This makes it plausible to assume that those philosophers who treat the question of the existence of numbers as a serious philosophical problem and offer lengthy arguments on either side, do not have in mind the internal question. And, indeed, if we were to ask them: 'Do you mean the question as to whether the framework of numbers, if we were to accept it, would be found to be empty or not?', they would probably reply: 'Not at all; we mean a question prior to the acceptance of the new framework'. They might try to explain what they mean by saying that it is a question of the ontological status of numbers; the question whether or not numbers have a certain metaphysical characteristic called reality (but a kind of ideal reality, different from the material reality of the thing world) or subsistence or status of 'independent entities'. Unfortunately, these philosophers have so far not given a formulation of their question in terms of the common scientific language. Therefore our judgment must be that they have not succeeded in giving to the external question and to the possible answers any cognitive content.

Nevertheless, we should, Carnap (1950, p. 208) says, be charitable, and take at least some of them to have a *pragmatic meaning* that we can somehow *guess*:

Those who raise the question of the reality of the thing world itself have perhaps in mind not a theoretical question, as their formulation seems to suggest, but rather a practical question, a matter of a practical decision concerning the structure of our language. We have to make the choice whether or not to accept and use the forms of expression in the framework in question.

In other words, the question whether to accept a linguistic framework is a *pragmatic* one. In Carnap's view, the practical answer should be determined by the *scientific fruitfulness* of adopting the framework.

In sum, Carnap's distinction between internal and external questions is relative to a *linguistic framework* in which the subject matter is investigated. Internal questions are those that the scientific theory is equipped to deal with, at least in principle, using its methods. External questions are those that so to speak *transcend* the framework. Myhill (1955, p. 62) puts it thus:

A question ... is internal relative to [a linguistic framework] T if the asker accepts T at the time of his asking, and is prepared to use T in order to obtain an answer; external otherwise, in particular if the question is part of a chain of reflections and discussions aimed at choosing between T and some rival theory.

One and the same sentence can express an internal or an external question. Take again the question 'Are there numbers?' From within the framework of mathematics, the answer is obviously yes. But if it is interpreted as a 'philosophical' question, the answer is not clear at all. We have seen that Carnap holds that the only clear way to interpret this question as an external one is by taking it as a question about whether the framework of mathematics is a good one to adopt. But this means that Carnap does not admit the *ontological* interpretation of the question 'Are there numbers?' From his empiricist views on meaning it follows that such questions are meaningless.

Few philosophers today subscribe to Carnap's version of meaning-theoretic empiricism, and therefore few are now willing to follow Carnap in classifying the question 'Are there numbers?' as meaningless. External questions are characterised as those questions that sound like *basic* internal questions but are not intended to be obvious, where *basic* internal questions are those that are *analytic*. But Quine (1951) has forcefully argued that the analytic/synthetic distinction is untenable. Many contemporary philosophers accept Quine's critique. Thus many philosophers take Carnap's distinction between internal and external questions to be problematic. After the demise of logical empiricism it is not easy to argue that the questions that Carnap classified as external have no 'representative' meaning.

Burgess is not a logical empiricist. In particular, he does not accept the logical empiricist theory of meaning. Nevertheless, he believes that Carnap was on to something. He believes that something like Carnap's distinction between internal and external questions survives the downfall of logical empiricism.

Consider what Carnap called basic *internal* questions. Burgess starts with a key point that was repeatedly stressed by Parsons: the answers to such questions are, from the point of view of the theory, *obvious*. This obviousness is not easily explained without invoking *some* notion of analyticity.

Burgess therefore proposes a new philosophical explication of analyticity. He calls Carnap's basic internal questions *pragmatically analytic* (Burgess 2005b). A question is pragmatically analytic if it would be *helpful* for anyone who disagrees with the answer given by the theory to use a different term for some of the concepts in the question (or put subscripts on some of the terms) (p. 54):

My proposal is that the [statement] should be regarded as 'basic', as 'part of the meaning or concept attached to the term', when in case of disagreement over the [statement], it would be helpful for the minority or perhaps even both sides to stop using the term, or at least to attach some distinguishing modifier to it.

So the view is that disagreeing with the answer of the theory to a pragmatically analytic question is a sign that you are implicitly rejecting some of the concepts in the question and/or are implicitly replacing them by other concepts.

Unlike Carnap's notion of analyticity, Burgess' notion of *pragmatical analyticity* is a vague one. But it is intended nonetheless to be a useful one. It explains why the answers to what Carnap calls basic internal questions are *obvious*. Moreover, it explains why it is still legitimate to ask, with Carnap, what the meaning is of what Carnap called *external* questions (Burgess 2004a, p. 35):

Since I do not wish to claim that the absence of empirical meaning is tantamount to the absence of all meaning, where Carnap would put forward a categorical negative, 'These questions are meaningless', I only put forward a rhetorical question, 'What are these questions supposed to mean?' But I do agree with Carnap that the question of the Real *existence of mathematical entities does lack* empirical *meaning ... [W]e have absolutely no idea of what difference it would make to how things look; or rather, we have a very strong suspicion that it would make no difference at all.*

In other words, Burgess argues that Carnap has isolated a worry about the question of the 'real existence' of entities that can be divorced from the logical empiricist framework in which it was couched. The issue is not that there is no framework in which these questions can be posed. After all, ordinary English may be regarded as a framework in which they are posed. The problem is that this framework contains no resources for even beginning to resolve them. This makes us wonder about the meaning of these questions in the first place.

In another publication, Burgess (2004b, p. 34) qualifies this judgement somewhat:

One very traditional sort of way to try to make sense of the question of the ulti-mate metaphysical existence of numbers would be to turn the ontological ques-tion into a theological question: Did it or did it not happen, on one of the days of creation, that God said, 'Let there be numbers!' and there were numbers, and God saw the numbers, that they were good? According to Dummett, and according to Nietzsche – or my perspective on Nietzsche – this is the only way to make sense of questions of ontological metaphysics. The Carnapian claim that ontological metaphysics is meaningless is roughly equivalent to the conjunction of this Nietzsche-Dummett thesis, 'realism makes sense only on a theistic basis', with analytic atheism, the thesis that theological language is meaningless. Both these theses are highly controversial: analytic atheism was explicitly rejected even by outspoken agnostics like Russell, and the Nietzsche-Dummett thesis is rejected by many philosophers in Australia who regard themselves as simulta-neously 'realists' in some strong sense, and 'physicalists' in some sense equally strong.

Burgess believes that analytical theism is meaningful but false, but that something like the Nietzsche–Dummett thesis is true. There may be some tongue in cheek here. But the message seems to be that there is no *inter-esting* sense in which the realism question is meaningful: there is only the 'theological' sense.

Now that we have arrived at Nietzsche, we may as well press on to discuss the position of Heidegger. Burgess rhetorically asks: 'what do external onto-logical questions *mean*?' But there is more to metaphysics than ontology. So one might go on to ask: 'what do metaphysical statements in general mean?' Heidegger (1949) extends Burgess' thesis by claiming that *all* metaphysical statements are meaningful only within a theological context. The slogan is that philosophy equals metaphysics, and metaphysics equals *ontotheology*:

Because it makes be-ing as be-ing an idea, metaphysics in itself is in fact two-in-one: the truth of be-ing in the most general sense and in the highest sense. In its essence it is ontology, in the narrower [scholastic] sense, and theology. This onto-theological essence of authentic philosophy ... must indeed be accounted for by the way it brings [be-ing], that is, as [be-ing], out into the open. The theological character of ontology is not due so much to the fact that Greek metaphysics was later absorbed by Christian sacred theology and transformed by it. It is due more to the means by which be-ing as be-ing had disclosed itself [sich entborgen hat] from early on. That emergence of be-ing first made it possible for Greek philosophy to overpower Christian theology.

Moreover, Heidegger adds, metaphysics is over, it is 'done'. We must move on to something else. He even has a suggestion: we must learn to *think*.[2]

Related sentiments are expressed by Arthur Fine (not Kit Fine!) in the philosophy of science, where he attempts to 'deflate' the realism debate (Fine 1984a, 1984b). A question that was much debated in the 1980s is whether it is rational to take non-observable entities (properties and relations) that are postulated by our best scientific theories to exist. The realist says yes; the anti-realist says no. Arthur Fine takes both the realist and the anti-realist to be wrong. By adding to what science says ('Electrons really exist'; 'Electrons do not exist in reality') both the realist and the anti-realist inevitably distort science. Instead, we should let science speak for itself, and not add a metaphysical interpretation to it.

A somewhat similar conviction drives the work of a philosopher of science of a more fiery temperament: Paul Feyerabend. Much of his work is motivated by a deep-seated belief that all efforts to find rational structure in the evolution of scientific theories are in vain (Feyerabend 1975). The evolution of scientific theories is always so multi-faceted, fragmentary, and complex that any rational reconstruction of it must be fundamentally misguided. By holding scientific practice accountable to norms that are foreign and external to it, the philosopher goes wrong. Not only does scientific practice fail to live up to the norms but the disembodied norms themselves should be treated with suspicion anyway.

Like Burgess, Tait (1986) has sought to deflate the platonism/anti-platonism debate in the philosophy of mathematics. According to Tait, questions of existence of mathematical entities can only sensibly be asked and reasonably be answered from within mathematical frameworks that are ultimately of an axiomatic nature. If you are working in number theory, for instance, then you can ask whether there are prime numbers that have a given property. Such questions are then to be decided on purely mathematical grounds. Philosophers have a tendency to step outside the framework of mathematics and ask 'from the outside' whether mathematical objects really exist and whether mathematical propositions are really true. In asking such questions they are asking for supra-mathematical or metaphysical grounds for mathematical truth and existence claims. Tait

---

[2] We will not pursue Heidegger's views on what it means to think further here.

argues that it is hard to see how any sense can be made of such external questions. He rhetorically asks (1986, p. 68):

As a mathematical statement, the assertion that numbers exist is a triviality. What does it mean to regard it as a statement outside mathematics?

By deflating such questions, Tait (2001, Section 2) attempts to bring them back to where they belong, i.e. to mathematical practice itself:

[An] objection [against platonism] is that, even though the axioms [of mathematics] might be entirely consistent and in their consequences express, as far as they go, our conception of numbers, sets, and so forth, they could nevertheless be false or meaningless because the objective universe of numbers and sets is just not like that, or worse, does not even exist ... I want to save [the term 'platonism'] for the view that we can truthfully assert the existence of numbers and the like without explaining the assertion away as saying something else. Realism in this sense is the default position: when one believes that mathematics is meaningful and has, as one inevitably must, finally become convinced that mathematical propositions cannot be reduced to propositions about something else or about nothing at all, then one is a realist ... Let me refer to the more extreme position as superrealism. *An important consequence of superrealism, and, as I believe, a telling objection to it, is that it implies an alienation of truth in mathematics from what we actually do: mathematics becomes* speculative *in the sense that even the most elementary computations, deductions, and propositions must answer to a reality ... about which we could be wrong.*

Motivated by Quine's (1969) epistemological naturalism, Maddy (2007) argues that there ultimately is not much at stake in the debate between platonistic and anti-platonistic (such as fictionalist) interpretations of mathematics. There is no higher court of appeal for mathematical existence than mathematical practice itself, with its own norms and standards and methods. But mathematics itself is silent about whether it should be interpreted in a platonistic or in an anti-platonistic fashion. Therefore it is hard to see how we could every decide between realist and non-realist interpretations of mathematics. So perhaps there is not much between them.

All this does not amount to a 'no philosophy' attitude: it does not mean that there is no role for the philosophy of mathematics. But this movement does encourage philosophers to stay as close to the internal perspective as possible, and it is firmly opposed to the traditional metaphysical approach to questions in the philosophy of mathematics. These are *quietist* positions (in the sense of Fine 2001, Section 4) in the philosophy of mathematics.

Quietist positions in philosophy can be traced back to the work of Wittgenstein (1953). He argued throughout the second part of his career that when philosophers put forward *theories* about the meaning of language, about mental and epistemic phenomena, they inevitably go wrong. Wittgenstein strives consistently to take philosophers away from their philosophical theories and lead them back to what they knew before they produced their theories. He never applied his quietist outlook to mathematics or to science. Indeed, he was critical about science and outright revisionist about mathematical practice. But his followers have rightly taken science and mathematics to be part and parcel of our 'form of life' and have treated these subjects philosophically in the same way as the disciplines against which Wittgenstein directed his anti-theoretical considerations.

2.3 Naive Metaphysics and Foundational Metaphysics

Fine (2017b) has proposed a new approach to metaphysics. At the heart of his proposal is an *alternative* division of metaphysics into two sub-disciplines:

1 naive metaphysics
2 foundational metaphysics

Fine conducts his own metaphysical investigations according to this new approach, which we may call *Finean metaphysics*.

Naive metaphysics is the metaphysical investigation of how the nature of the world appears to us. In naive metaphysics we take how things (properties, relations) appear to us at face value: 'we need to restore ourselves to a state of innocence in which the metaphysical claims are seen to be about the subject-matter in question [...] and not about our relationship to that subject-matter' (Fine 2001, p. 7).

The object of naive metaphysics is not on the side of us. It should not be confused with the investigation of our *experience* of reality (investigation of 'sense data' or anything like it) or our construction of reality (if there is such a thing), as it might for instance be carried out in phenomenology. Nor should it be confused with an investigation of our *conception* of reality, as in Strawson's (1959) descriptive metaphysics.

The activity of doing naive metaphysics is fundamentally constructive and creative, somewhat like the activity of doing mathematics is. It involves an *openness* to metaphysical phenomena and a lack of preconceptions.

Naive metaphysics is by no means an unconstrained activity. Not everything goes: it can be done exceptionally well, it can be done extremely poorly, and anything in between. But it is done without suspicions about the source of the appearances. So it is anti-sceptical, and anti-critical if critical is taken in the Kantian sense of the word. Hence, I suppose, the label 'naive'.

Naive metaphysics tries to describe how reality *directly* metaphysically appears to us. This is very different from *abductive reasoning* ('inference to the best explanation', 'success arguments'), which to a large extent governs the ontological stage in much of contemporary metaphysics. It also means that naive metaphysics is not much interested in ontological reduction.

There is not much room for problems of *scepticism* in naive metaphysics. Nonetheless, naive metaphysics can, in Fine's view, be done well and still go wrong: there is a genuine possibility that how the nature of some things appears to us is not how those things really are. Part of Appearance may be an *illusion*.

The thought of a distinction between how matters appear and how matters stand is fuelled by intuitions originating from our everyday contact with the concrete world around us and is sharpened by empirical science. The stick appears bent, but when we pull it out of the water, it turns out to be straight. The outwards-directed upper horizontal double arrow appears shorter than the lower inward-directed horizontal arrow, but when we measure them, it turns out that they are of equal size (the Müller–Lyer illusion). Gödel (1947) famously suggested that as it is in ordinary life and in science, so it is in mathematics. Unrestricted comprehension for sets appears to be obviously true – it did appear so for Frege. But when we analyse the situation carefully, we come to realise that comprehension for a predicate must be restricted to an antecedently given set.

This notion of a distinction between appearance and reality can then also be applied, in Fine's view, to metaphysics. So the second part of metaphysics, foundational metaphysics, addresses whether and to what extent the way things appear to be accords with how things are in reality. In this way, foundational metaphysics is closer to the ontological part of what Fine calls traditional metaphysics.

Foundational metaphysics is critical by nature, and it is very much interested in ontological reduction: it is anything but naive. A foundational metaphysical investigation may also reveal that *nothing* can satisfy the metaphysical constraints that are uncovered by the prior naive metaphysical inquiry. Thus Fine's division between naive and foundational metaphysics differs from the bipartite division in the Meinongian way of doing metaphysics: even if a kind of object is revealed in the second stage of a Meinongian investigation not to exist, it will still be guaranteed to 'subsist'.

Fine argues that it is of essential importance that naive metaphysics precedes foundational metaphysics. If we embark on an investigation in foundational metaphysics on the strength of existence statements obtained by some naturalistic method instead of going *first* through a thorough exercise in naive metaphysics, then we are in danger of losing sight of important metaphysical phenomena (Fine 2017b, Section 2). If we jump immediately to some reductive proposal, for instance, then important metaphysical constraints that adequate accounts have to satisfy do not even get onto the agenda. Suppose, for instance, that we are conducting a metaphysical investigation into tables. If we immediately conceive of tables as Eddington proposed (i.e. as an arrangement of molecules in a volume of mostly empty space), then we are discouraged from inquiring with an open mind into persistence conditions for tables, and counterfactual scenarios involving tables are not judged without preconceptions.

2.4 Naive Metaphysics Only

While we are doing naive metaphysics in Fine's sense, we are *suspending* the question whether the nature(s) that we are investigating also exist in reality. What I will now suggest is that it is not clear that the question from foundational metaphysics whether this nature really exists, makes any sense. In other words, I will suggest that there may be no role for foundational metaphysics to play.

Fine tries to make room for foundational metaphysics by drawing a distinction between *what is apparently the case* and *what is really the case* (Fine 2001, Section 1). But given that we have to think of appearance in this context as metaphysical appearance (Section 2.3), many philosophers find this distinction somewhat obscure (see e.g. Szubka, 2016). Let us see what this distinction can amount to, and whether it holds water.

I will proceed on the assumption that the quietist critique of ontology of science, and specifically of physics (Carnap, A. Fine) and of mathematics (Maddy, Tait), is substantially correct.

Physics has at times articulated and entertained *coherent* theories about kinds of entities and properties that turned out not to exist: magnetic monopoles, the aether, vital forces, etc. So there is a sense *internal to physics* in which there can be matters that appear to be the case but are not the case in reality.

It is not clear that the same has occurred in the history of mathematics. Certainly kinds of entities have been entertained that turned out to be incoherent – Reinhardt cardinals, and indeed the set of all non-self-membered sets, are cases in point. But when set theorists are convinced that a certain type of large cardinals *could exist*, then they take them to in fact exist. (This phenomenon is connected to the remarkable phenomenon that the large cardinal axioms seem to line up linearly in consistency strength.) A better way to put this is perhaps to say that the distinction between possibility and existence seems to *dissolve* in set theory, and of course even more so in the rest of mathematics. And this is so despite the fact that it would not be inconsistent to maintain that some sets may consistently be taken to exist even though they do not exist 'in reality'. Again, all of this is *internal* to mathematical practice.

So even if we adopt a quietist view regarding the metaphysics of physics and of mathematics, there still is an important *internal* difference between them. This difference is connected with a methodological difference between these two disciplines. Theoretical physics is governed to a large extent by inference to the best explanation of the empirical phenomena. In mathematics, the situation is subtly different. Gödel (1947) emphasised that inference to the best explanation also plays a role in mathematics, viz. in the search for new axioms. This may be true. But I think, *pace Gödel*, that it does so in a way that is different from physics. Giving a good explanation of phenomena lower down (properties of the real numbers, for instance) constitutes evidence for the *coherence* of a high level new axiom with what we already know. And coherence and existence, from the perspective internal to mathematics, amount to the same.

According to the prevailing view since the early days of analytical philosophy, metaphysics is closer to physics than to mathematics in this regard: there is a gap between coherent appearance and reality. Russell (1919, Chapter 16) expresses this sentiment in the following piece of foundational metaphysics of mathematics:

It is argued, e.g. by Meinong, that we can speak about 'the golden mountain', 'the round square', and so on; we can make true propositions of which these are the subjects; hence they must have some kind of logical being, since otherwise the propositions in which they occur would be meaningless. In such theories, it seems to me, there is a failure of that feeling for reality which ought to be preserved even in the most abstract studies. Logic, I should maintain, must no more admit a unicorn than zoology can; for logic is concerned with the real world just as truly as zoology, though with its more abstract and general features. To say that unicorns have an existence in heraldry, or in literature, or in imagination, is a most pitiful and paltry evasion. What exists in heraldry is not an animal, made of flesh and blood, moving and breathing of its own initiative. What exists is a picture, or a description in words. Similarly, to maintain that Hamlet, for example, exists in his own world, namely, in the world of Shakespeare's imagination, just as truly as (say) Napoleon existed in the ordinary world, is to say something deliberately confusing, or else confused to a degree which is scarcely credible. There is only one world, the 'real' world: Shakespeare's imagination is part of it, and the thoughts that he had in writing Hamlet are real. So are the thoughts that we have in reading the play. But it is of the very essence of fiction that only the thoughts, feelings, etc., in Shakespeare and his readers are real, and that there is not, in addition to them, an objective Hamlet. When you have taken account of all the feelings roused by Napoleon in writers and readers of history, you have not touched the actual man; but in the case of Hamlet you have come to the end of him ... If no one thought about Hamlet, there would be nothing left of him; if no one had thought about Napoleon, he would have soon seen to it that some one did.

This is still the received view in metaphysics today. And in contemporary metaphysics, this is connected, at least since Quine, with *the abductive method*. David Lewis, the most influential metaphysician of the past four decades, followed the methodological *dictum* that the metaphysical theory that best explains the given is rationally to be preferred over all others, where the given consists of our best scientific theories, common sense, folk psychology, empirical experience, philosophical intuitions, etc. A similar methodological line is taken by leading contemporary metaphysicians such as Sider (2011, p. 12) and Williamson (2013, Methodological Afterword, pp. 423–429).

Lewis (1986, p. 4) describes his methodology as follows:

We have only to believe in the vast range of possibilia, *and there we find what we need to advance our endeavours. We find the wherewithal to reduce the diversity of notions we must accept as primitive, and thereby improve the unity and economy of the theory that is our professional concern – total theory, the whole of what we take to be true. What price paradise? If we want the theoretical benefits that talk of* possibilia *brings, the most straightforward way to gain honest title to them is to accept such talk as literal truth ... The benefits are worth their ontological cost. Modal realism is fruitful; that gives us good reason to believe that it is true.*

Since perfect harmony between the theory and the given is completely unachievable, a balance must be struck: such is the philosophy of the accountant. The motto is that the simplest and most powerful metaphysical theory wins – and Lewis recognises of course that in practice all victories in this game are temporary. But this opens up a space between how things metaphysically appear to be the case and how things really are: we are in a similar situation to the one we are in in physics. I disagree with this view of how metaphysics should be done, and will now sketch an alternative.

Metaphysics is not an interpretation of the subject matter of other practices (common sense, folk psychology, empirical science, literature, mathematics). Rather, it is a self-standing discipline alongside the empirical sciences and mathematics, even though it can take physical entities (such as material objects) or mathematical objects (such as natural numbers) as objects of metaphysical theorising. So, in accordance with quietism, I say that the metaphysics of mathematical entities should not be in the business of giving an answer to the question what mathematics is really about. For this reason I regard the slogan that 'mathematics is the science of structure' (see Section 1.1) with suspicion.

If we take this line, then we can distinguish between internal and external questions *of metaphysics*. This is a Carnapian move, even though Carnap himself would strongly disapprove of making it.

The internal questions of metaphysics are addressed by naive metaphysics. As we have seen, naive metaphysics investigates the metaphysical phenomena. Like the empirical sciences and mathematics, naive metaphysics has its own rules and methods, norms and standards, etcetera. Metaphysicians can distinguish between good and bad metaphysics, and they do this every day, even though perhaps any two metaphysicians disagree about the quality of some pieces of metaphysics.

One of the regulative principles of naive metaphysics is *consistency*. When metaphysical appearances contradict each other – as they sometimes do – we feel a rational urge to resolve the contradiction. Should not this urge be explained by a conviction that *how matters really stand* cannot be contradictory?[3]

The naive metaphysician does not *have* to appeal to an ultimate metaphysical reality to explain the methodological maxim of consistency. Perhaps consistency is so basic a norm of thought that it cannot be explained in terms of anything more fundamental. But our discussion does show that

---

[3] Thanks to Gideon Rosen for raising this objection.

naive metaphysics aims at more than merely a *description* of metaphysical appearance. Instead, its goal is to arrive at an *understanding* of metaphysical appearance. To arrive at understanding, we use logic.

Let us now turn to the question which parts of good metaphysics express how matters stand in reality and which parts fail to do this. In other words, what about *foundational metaphysics*?

On this question, I take a quietist stance. If metaphysics as an intellectual project stands on its own feet alongside the empirical sciences and mathematics, then the quietist critique applies as much to metaphysics as it does to the empirical sciences and mathematics. A Carnapian distinction between internal and external questions can be made *inside* metaphysics, and it is not clear that external questions about naive metaphysics make sense. Asking which parts of our good metaphysical theories correspond to how matters really stand is like asking which parts of mathematics correspond to how matters really stand in the mathematical world.

Perhaps we should do away with the part of foundational metaphysics that is concerned with real existence; perhaps metaphysics is not in need of the external validation that foundational metaphysics promises. Instead of trying to scramble to a higher meta-metaphysical ground, let naive metaphysics speak for itself and let us leave it at that.

In this way, metaphysics is like mathematics. Matters appear to be complicated somewhat by the fact that naive metaphysical theories often seem to contradict each other: this is almost never the case in mathematics. I think that too much is sometimes made of this contrast. Where two good metaphysical theories contradict each other, they often both contain important insights into the metaphysical phenomena under consideration.

This does not mean that there is no room for fundamentally criticising a naive metaphysical enquiry. Suppose that a metaphysician enquires into the nature of *F*s. Then an *F*-sceptic can ask pointed questions about the nature of *F*s. The more it seems that the naive metaphysician has no means of answering the sceptic's questions in reasonable ways, the more her enquiry is fundamentally undermined.

Moreover, in contrast to questions of real existence, there may be room for questions of reduction *inside* naive metaphysics.[4] After all, it cannot be excluded that reductions can shed light on the *nature* of properties and relations.

We have seen that for Fine, matters of reduction belong to foundational metaphysics rather than to naive metaphysics. So I am carving 'Finean'

---

[4] We will see an example of this in Section 7.2.

metaphysics in a slightly different way from the way Fine is carving it. Nonetheless, given the poor track record of grand programmes of ontological reduction in metaphysics, strong reductionist claims should be treated with caution.[5]

The best and most interesting part of metaphysics does not lie in foundational metaphysics. Questions of realism often generate more heat than light and are often of less import than they are taken to be. Consider, for instance, the ontological debate between Lewis and Kripke. They take opposite stances on questions of *foundational* metaphysics of modality, i.e. questions related to whether and in what sense beside the actual world there *really exist* other possible worlds. But the true interest of possible world theory lies, in my view, in how it figures in theories of essence and of natural kinds, of reference, of kinds of conditionals. If Lewis and Kripke put aside their foundational metaphysical disagreements for a moment – they have done so on countless occasions – then they can agree on many matters in the naive metaphysics of modality. Surely, for instance, Lewis can appreciate the deep insight contained in the Kripke–Putnam theory of natural kinds.

It is not easy to say what makes for a good piece of metaphysics, just as it is not easy to say, for example, what makes something a good piece of mathematics. Good metaphysics often relates what appear to be different parts of metaphysics in deep and surprising ways. Good metaphysics is often fruitful: it has interesting applications to problems in philosophy that it was not originally intended to shed light on. It often explains why other metaphysical views have gone wrong in the way that they did.

So naive metaphysicians can disagree over how the nature of a kind of objects (or properties, or relations) appears to us. We do not perceive metaphysical natures with perfect clarity, and we sometimes misdescribe them. I hope we can gradually correct each other's mistakes, and thus make metaphysical progress. But I know of no pre-described *method* for doing this systematically. In Section 11.1, when looking back on what has been accomplished in this book, I will address the question whether there are at least some *criteria* that can help us to distinguish good metaphysics from bad metaphysics.

---

[5] I suspect that *grounding* would play an important role in Fine's full account of what foundational metaphysics should look like. I will not go into that here. A good article on metaphysical dependence and grounding is Rosen (2010).

2.5 Metaphysics of Mathematics

If metaphysics and mathematics stand alongside each other as cognitive endeavours, if metaphysics of mathematics is part of metaphysics, and if we should take a quietist stance towards questions of foundational metaphysics, then how does metaphysics of mathematics differ from mathematics itself?

Consider, to take a typical example, arithmetic. Here is a rough and non-exhaustive answer *internal* to mathematical practice to the question of the subject matter of arithmetic: *arithmetic is the investigation of the natural numbers* $0, 1, 2, \ldots$, *the properties that these natural numbers have, and the relations in which they stand to each other.* This may not be all that arithmetic is about. Nonetheless, I take this answer to be in the right direction. I take this answer to be fairly unexceptionable, but also metaphysically unrevealing.

The *metaphysics* of mathematics is a very different activity. It starts from mathematical practice in the widest sense of the word. This encompasses what is written in mathematical journals, what mathematics teachers say when they teach, what mathematicians say when they give research seminars, how ordinary people engage with and make use of mathematics in their daily lives, how empirical scientists think about mathematics, how engineers use mathematics and think about it, what we read in articles about mathematics in popular science magazines. We find that in mathematical practice, particular *concepts* are used: computation, set, space, continuum, number, finite, infinitely large, infinitely small, function, arbitrary (!), proof, structure, diagram, and so on. We naively take these concepts to refer to entities, properties, and relations. It is the task of the metaphysician to uncover the *nature* of these entities, properties, and relations. So this gives rise to metaphysical research which results in metaphysical theories. This research is regulated by norms, heuristics, and rules that are intrinsic to metaphysics.

I will do metaphysics of mathematics in a naive and 'innocent' way. Questions of existence are bracketed, except to the extent that they are internal – like existence questions typically are in mathematics. I forgo the claim that metaphysics tells us what mathematics, or parts of mathematics, are *really* about.

Fine, like many other philosophers, takes it to be a task of the metaphysician to offer us a *worldview*. Metaphysics should show us how it all hangs together (Fine 2001, pp. 1–2):

Among the most important issues in philosophy are those concerning the reality of this or that feature of the world. Are there numbers or other abstract objects? Is everything mental or everything physical? Are there moral facts? It is through attempting to resolve such questions that philosophy holds out the promise of presenting us with a worldview, a picture of how the world is and of our place within it.

So metaphysics should tell us, in Fine's view, what mathematics is about and how what it is about is related to the rest of the world.

Against this, I say that metaphysics is under no such obligation. Metaphysics has no *representative* function; it does not have to give us a picture of the world. It has been immensely fruitful to look for deep connections between seemingly unrelated metaphysical domains. But looking for an overall metaphysical worldview has, in my opinion, always been stifling and crippling: it has tended to blind philosophers to metaphysical phenomena. Moreover, it seems wildly optimistic to assume – certainly at this stage – that there is an overall worldview to be had: that everything does hang together in any informative way. At any rate, I do not have a worldview to offer.

Someone might object that surely I want the result of my metaphysical investigation to be *true*, and to be true *is* to say how matters really stand. But I do not accept that the conclusion follows. I indeed hope to arrive at a metaphysical view about mathematical structures that is to a large extent true. But the conclusion of the objector presupposes that some version of the correspondence theory of truth is correct. This is not the place to have this discussion, but I reject this presupposition. Instead, I have long been and still am a card-carrying deflationist about truth (Horsten 2009).

The quietist position that I advocate should be distinguished from Schiffer's view according to which ontological conclusions can be 'easily' obtained from what he calls *something-from-nothing* arguments (Schiffer 2003). According to the latter view, from the statement

This chestnut is brown.

we can easily infer the ontological statement

There are properties.

via a something-from-nothing argument. The ontological statement does not follow logically from the innocent common sense judgement. Nonetheless, the latter is *conceptually entailed* by the former: using semantical rules that are definitional of the concept of property, the latter statement can be obtained from the former.

Unfortunately, such something-from-nothing arguments are dialectically ineffective. Sider (2011, Section 9.9) is right when he argues that even if the rule that is used is 'definitional' of the concept 'property', it may nonetheless not be truth-preserving. However, the ontological quietist is not concerned by this: she was not in the business of giving arguments for ontological conclusions in the first place.

2.6 An Objection from Pseudo-science

> Be kind to everyone on the way up;
> you'll meet the same people on the way down.
>
> *(ascribed to Wilson Mizner, 1932)*

Quine argued on the basis of confirmational holism that we should take all mathematics that plays a significant role in our best scientific theories of the world to be part of our overall best theory of the world. Moreover, Quine (1984, p. 788) stated that we may round off this amount of mathematics to an elegant, compact and coherent whole. Perhaps the mathematics that plays an important role in empirical science is the mathematics that describes objects that have set-theoretical representations in $V_{\omega+4}$. But the theory of $V_{\omega+4}$ cannot be concisely and perspicuously stated, so we may in a liberal mood round 'legitimate' mathematics off to ZFC. Large cardinal theory is then relegated to the light entertainment section and should not be taken ontologically seriously.

Maddy (1997) resists Quine's argument. She argues that just as empirical science does not answer to a higher tribunal, neither does mathematics. Mathematics does not derive its legitimacy as a body of knowledge from its contributions to empirical science. (This is of course perfectly compatible with the fact that much energy and resources go to parts of mathematics that have direct or indirect applications in science.) The internal methods and standards of mathematics as a shared practice determine what counts as mathematical knowledge and what does not.

I propose that we do for metaphysics what Maddy proposed to do for mathematics. Empirical science has its own internal standards to judge empirical theories by; mathematics has its own internal standards for success; *so does metaphysics*. An objection to this line of reasoning was raised by Maddy herself (Maddy 1997, pp. 203–205) and was further explored by other authors (Dieterle 1999; Antonutti Marfori 2012).

Astrology too has its own internal standards and measures for success. So does speculative theology. Are we then to count these disciplines as bodies of knowledge alongside empirical science and mathematics?

Concerning astrology, Maddy points out that, in contrast with empirical science, the predictive power of astrology has consistently been weak. Concerning speculative theology, she points out that, unlike mathematics, theology is not integrated in empirical science. So astrology is a *pseudo-science*, and theology is not entitled to claim scientific status. But then again, metaphysics is in this respect on a par with theology. So if Maddy's response is right, then I cannot claim a similar status for metaphysics as she has claimed for mathematics.

Maddy's argument has been criticised by Dieterle (1999) and Antonutti Marfori (2012). The problem, as the authors see it, is that Maddy seems to appeal implicitly to confirmational holism in her response. But then she opens herself to the charge that, for instance, the higher reaches of set theory, are not integrated in empirical science. And if she forgoes any appeal to support for mathematics from empirical science, then astrology and theology are on a par with mathematics (Dieterle 1999, p. 133).

Perhaps Maddy has been misunderstood. Astrology, as it has actually been practiced, cannot be treated as *totally* separate from science. Science is in the business of empirical predications; some even say that making successful predictions is the aim of science. Astrology has always made empirical predictions: this is among its core objectives. Therefore astrology deserves to the measured by the method of science. Sadly, it has failed the test miserably.

Speculative theology is a more complicated matter. It does not make empirical predictions. In our days it is regarded with much suspicion, whereas in the Middle Ages it was regarded as the queen of the sciences. (But mind the epigraph to this section.) At the time when Western theology was at its peak, it was taken to describe phenomena that would be as absurd to deny as the existence of the external world. When Anselm (1977, Chapter 1) wrote, approvingly quoting the Bible, that 'the fool in his heart says "there is no God" ', he meant to be taken literally. One can deny the existence of God hypothetically, in the context of an intellectual exercise, to see if a contradiction can be derived from this assumption. But a sane person, Anselm contends, cannot really *believe* that there is no God. The problem with theology today, I think, is that the existence of God no longer forces itself upon most of us: 'God is dead', said Nietzsche. So it is no longer clear that speculative theology answers to a shared phenomenon or appearance.

But this is where theology differs from metaphysics. There *are* ways in which things metaphysically appear to us. Perhaps these ways in which things metaphysically appear to us will likewise evaporate at some point in the future. Perhaps the day will even come when it will no longer appear to us that there are other *persons*, but just biological bodies. Cross that bridge when we get there.

In sum, just as mathematics answers to phenomena, metaphysics does so too. Mathematics needs no ontological support from empirical science; neither does metaphysics.

2.7 Mathematical Models

Naive metaphysics is interested in the metaphysical phenomena. It focuses on a kind *F* of entities and investigates how the nature of the *F*s, the nature of their properties, and the nature of the relations in which they stand to each other and appear to us.

The *F*s typically do not metaphysically appear to us as constituting *mathematical* entities of any kind. Moreover, the naive metaphysician is not interested in the question whether the way in which the nature of the *F*s appears to us can somehow be *reduced* to some mathematical entity. Yet *mathematical models* will turn out to play a major role in this book. In particular, I will model arbitrary objects and generic systems as *sets*. (I regard it as an open question of some importance whether some of the metaphysical phenomena that I will investigate could be better modelled in category theory, for instance.)

In the natural sciences, models can be valuable even if they are fundamentally unrealistic, not in the sense of making idealisations (such as absence of friction suppositions in mechanics), but in the sense of *intentionally* making fundamentally false assumptions – as is done for instance in modelling traffic as a fluid passing through a system of connected tubes (Horsten 2013, Section 5.2). Even though such models do not really explain anything, they serve an important goal: they are connected to observational and experimental predictions. Even models that do not describe the world anywhere near correctly can be extremely powerful as a source of empirical predictions. Indeed, even an empiricist such as van Fraassen, who is agnostic about the existence of unobservable entities, properties, and relations, is happy to acknowledge the value of models that postulate sub-atomic particles. An instrumentalist stance to models is always possible in science.

Mathematical models are only of use to the naive metaphysician if they capture the *structure* of the nature of things as they metaphysically appear to us. The debate whether a given model or class of models succeeds in doing this, must of course be conducted in the philosophical style. And this is to a large extent where the philosophical action is. In this sense, the discursive style necessarily forms a constitutive part of the metaphysical investigation.

Abductive reasoning does not play a role here. It is not because a model or class of models explains the relevant metaphysical phenomena best that they should be accepted, or at least taken seriously. The models are not intended to *explain* anything. What they are meant to do is mathematically to *describe*, at the right level of detail, the structure of the nature (as it appears to us) of the entities we are interested in.

The naive metaphysician aims at uncovering *general metaphysical principles* governing the kinds of entities that she is investigating. A presentation of the metaphysical phenomena in ordinary language is apt to be to some extent vague and to display a certain fluidity, which makes these general metaphysical principles hard to discern. Provided that they capture the structure of the relevant metaphysical phenomena, mathematical models do achieve the *precision* that is needed to uncover the general principles that the naive metaphysician is searching for Horsten (2013, Section 5.4). By analysing such a class of models, we come to understand better the *content* of the metaphysical phenomena that we are interested in. Nonetheless, the mathematical models should never be confused with the metaphysical theory itself.

Some of the metaphysical principles that are made true by mathematical models that capture the structure of the metaphysical nature of *mathematical structures* may have a claim to being *axioms* of a theory of mathematical structure. But this must not lead us to infer that we are in the business of building a *mathematical* theory of mathematical structure. Set theory and category theory are *mathematical* theories of mathematical structure; the theory that I seek to develop in no way pretends to one day develop into a rival to them. Structure is a philosophical concept: mathematical structures are metaphysical entities. Set theory and category theory aim to capture aspects of the notion of mathematical structure that can play mathematical roles; I aim to contribute to our understanding of the metaphysical nature of mathematical structures.

3 | Arbitrary Objects

Schoolmaster: 'Suppose x is the number of sheep in the problem'.
Pupil: 'But, Sir, suppose x is not the number of sheep'.

(Littlewood [1986, p. 59])

I will now introduce the notion of arbitrary object and give an initial discussion of it. My discussion will be in the spirit of naive metaphysics, so foundational critique and questions of ontological reduction will take a back seat. Instead, I will take puzzles surrounding the notion of arbitrary object as key questions that motivate and inspire the construction of a metaphysical theory of arbitrary objects.

It will come as no surprise that there is considerable overlap between the theory of arbitrary objects that I advocate and that of Kit Fine. We will also see how basic elements of the theory of arbitrary objects can already be found in the theory of variables in Russell's *Principles of Mathematics*. Nonetheless, in this chapter I concentrate on my own view of the nature of arbitrary objects. The focus will be on giving an initial *exposition* of my account: only brief discussions of some immediate objections will be given. Fine's theory of arbitrary objects will be discussed in Chapter 7. A proper defence of my view is postponed until later chapters.

3.1 The Motivating Idea

Consider the following sentences:

1 The man on the Clapham omnibus likes fish and chips.
2 The cow is a mammal.
3 Take an arbitrary natural number: either it is 0 or it is larger than 0.

These sentences raise the question:

Question 3.1. *What do the expressions 'the man on the Clapham omnibus', 'the cow', 'an arbitrary natural number' stand for?*

The received view on question 3.1 is eloquently expressed in Russell's (1905) *On Denoting* and is known to every first year philosophy student.

On this account, these expressions do not stand for objects at all, but only appear to do so. This is seen when one inspects the *logical form* of sentences containing these descriptions, which gives the meaning of sentences in which definite and indefinite descriptions occur. The logical form of sentence 2, for instance, is

$$\forall x : Cx \rightarrow Mx.$$

So the subject-predicate form of sentence 2, which suggests that the expression 'the cow' denotes an object, is misleading: on closer inspection, it is a quantificational expression in which no complex denoting name occurs.[1]

Frege (1960, p. 109) suggested an alternative answer to question 3.1:

Of course we may speak of indefiniteness here; but here the word 'indefinite' is not an adjective of 'number', but ['indefinitely'] is an adverb, e.g., of the verb 'to indicate'. We cannot say that n designates an indefinite number, but we can say that it indicates numbers indefinitely. And so it is always when letters are used in arithmetic, except for the few cases (π, e, i) where they occur as proper names; but they designate definite, invariable numbers.

Breckenridge and Magidor (2012) defend Frege's view. They argue that the expressions 'the man on the Clapham omnibus', 'the cow', 'an arbitrary natural number' refer to ordinary, specific objects, but it is in principle impossible for anyone to know to which one they refer. In other words, some kind of *arbitrariness* is attributed to the reference relation for these expressions. We will come back to Breckenridge and Magidor's view, and issues surrounding it, later in this book (Section 8.5).

A third answer to question 3.1 is discussed and elaborated by Kit Fine (1983, 1985b). On this view, there are not only specific men, but also *arbitrary* men. Similarly, there are not only specific cows (such as Bella) and specific natural numbers (0, 1, 2, 3, . . .), but also arbitrary cows and arbitrary natural numbers. The expression 'the man on the Clapham omnibus' can be taken to refer to the arbitrary man (or perhaps the arbitrary British man), the expression 'the cow' in sentence 2 can be taken to refer to the arbitrary cow, the expression 'an arbitrary natural number' in sentence 3 can be taken to refer to the arbitrary natural number.

The first answer approaches questions about arbitrary objects in an indirect, semantic way. Naive metaphysics encourages us to approach the question more innocently, more directly. Even if Frege and Russell are right in their explication of the logical form of sentences 1–3, then we can ignore

---

[1] In linguistics and in philosophy of language, this naive Russellian account of the semantics of generics has come under pressure: see *Nickel* (2017), for instance.

natural language semantics and *still* ask the question what arbitrary objects are like.

The concept of arbitrary object is not a new idea. It was used frequently and mostly uncritically in mathematics until the time of Frege, who strongly objected to it. After Frege's critique, the idea of arbitrary objects was generally not considered to be viable until Kit Fine resuscitated it in the 1980s. But even after his work the idea had at most a handful of supporters. As a matter of fact, Fine (1983, p. 57) himself did not endorse the existence of arbitrary objects. He saw them as probably ontologically reducible to less controversial entities:

> *If now I am asked whether there are arbitrary objects, I will answer according to the intended sense of 'there are'. If it is the ontologically significant sense, then I am happy to agree with my opponent and say 'no'. I have a sufficiently robust sense of reality not to want to people my world with arbitrary numbers or arbitrary men. Indeed, I may be sufficiently robust not even to want individual numbers or individual men in my world. But if the intended sense is ontologically neutral, then my answer is a decided 'yes'. I have, it seems to me, as much reason to affirm that there are arbitrary objects in this sense as the nominalist has to affirm that there are numbers.*

As a naive metaphysician, I will not be concerned with ontological reducibility of arbitrary objects to anything else. Moreover, I do not understand what the expression 'the ontologically significant sense of existence' means. Instead, I will join Fine in trying to understand what arbitrary objects are like.

It is very doubtful that the concepts 'arbitrary', 'typical', 'average', 'variable', 'paradigmatic', and 'generic' all have the same semantic content. So I will restrict myself to using the term 'arbitrary', and the term 'generic', where I regard the latter as a term of art with the same meaning as 'arbitrary'. And arbitrary objects are to be contrasted with *specific* objects (Fine 1983, p. 63, calls the latter *individual* objects), which are the non-arbitrary objects of some accepted background ontology.

3.2 Early Russell on Variables

What is a variable? Today this is not seen as a pressing philosophical question, not because *variable* is not a philosophically important concept but because it is very generally accepted that we know the answer to the question.

The received view, in a nutshell, is this. Variables are *symbols*. In contrast with constant symbols, which have a fixed denotation, variables do not have a fixed denotation. Therefore also an expression in which a variable occurs, such as

x is a horse,

is an *incomplete expression*: it has no fixed meaning. This view was held by Frege, but it was also accepted by Russell. Indeed, it is a cornerstone of his logical analysis of indefinite and definite descriptions in *On Denoting* (Russell 1905). On this view, despite superficial appearances to the contrary, an indefinite description such as 'a woman', or even a definite description such as 'the prime minister of the United Kingdom in 2018', do not refer. The apparatus of quantification and binding of variables is used to show that, nonetheless, sentences involving indefinite and definite descriptions can be given determinate (and reasonable) truth conditions.

But during a period before that, which lasted until less than a year before the publication of *On Denoting*, Russell held a quite different theory of variables. In this section, I want to consider the theory of variables of Russell's *Principles of Mathematics* (Russell 1903b) in some detail.

According to the view expressed in the *Principles of Mathematics*, whatever we can think of is called a *term*, where the word 'term' can be taken to be synonymous with the term 'entity' (Russell 1903b, p. 53). The terms are then divided into *things*, on the one hand, and *concepts*, on the other hand (pp. 54–55). Things are terms that always 'occur as subject' in a proposition, whereas concepts can also occur in other ways. Notably, concepts can occur as predicates. Concepts are entities that, in contradistinction with things, *denote* (p. 53).

Early in his book, Russell (1903b, pp. 58–59) gives an initial characterisation of the differences in denotation of concepts expressed (in the context of a proposition) by expressions of the form 'all *A*', 'every *A*', and 'any *A*':

All *a*'s, *to begin with, denotes a numerical conjunction; it is definite as soon as a is given. The concept* all *a*'s *is a perfectly definite single concept, which denotes the terms [i.e., entities] of a taken all together. The terms so taken have a number, which may thus be regarded, if we choose, as a property of the class-concept [a], since it is determinate for any given class-concept.* Every a, *on the contrary, though it still denotes all the a's, denotes them in a different way,* i.e. *severally instead of collectively. Any a denotes only one a, but it is wholly irrelevant which it denotes, and what is said will be equally true whichever it may be. Moreover,* any *a denotes a variable a, that is, whatever particular a we may fasten upon, it is certain that* any *a does not denote that one; and yet of that one any proposition is true which is true of any a.*

This view of the denotation of *any a* is intimately connected with the theory of variable that Russell (1903b, Chapter VIII) formulates. It is worth quoting Russell somewhat at length here (pp. 90–91):

Originally, no doubt, the variable was conceived dynamically, as something which changed with the lapse of time, or, as is said, as something which successively assumed all the values of a class. This view cannot be too soon dismissed. If a theorem is proved concerning n, it must not be supposed that n is a kind of arithmetical Proteus, which is 1 on Sundays and 2 on Mondays, and so on. Nor must it be supposed that n simultaneously assumes all its values. If n stands for any integer, we cannot say that n is 1, nor yet that n is 2, nor yet that it is any other particular number. In fact, n just denotes any *number, and this is something quite distinct from each and all of the numbers. It is not true that 1 is any number, though it is true that whatever holds of any number holds of 1 . . .*

We may distinguish what may be called the true or formal variable from the restricted variable. Any term [i.e. any entity] *is a concept denoting the true variable; if u be a class not containing all terms, any u denotes a restricted variable. The terms included in the object denoted by the defining concept of a variable are called the* values *of the variable: thus every value of a variable is a constant. There is a certain difficulty about such propositions as 'any number is a number'. . . . If 'any number' be taken to be a definite object, it is plain that it is not identical with 1 or 2 or 3 or any number that may be mentioned. Yet these are all the numbers there are, so that 'any number' cannot be a number at all. The fact is that 'any number' denotes one number, but not a particular one . . .*

The notion of restricted variable can be avoided . . . by the introduction of a suitable hypothesis, namely the hypothesis expressing the restriction itself . . . By making our x always an unrestricted variable, we can speak of the *variable, which is conceptually identical in Logic, Arithmetic, Geometry, and all the other formal subjects. The* terms *dealt with are always* all *terms; the complex concepts that occur distinguish the various branches of Mathematics.*

Fine (1983, p. 55; 1985b, p. 5) takes this passage to be a critique of the notion of arbitrary object. But the contrary is the case. The passage strongly suggests that on Russell's view in *Principles of Mathematics*, variables are arbitrary objects in something like Fine's sense!

In fact, Russell's view concerning the denotation of the concept expressed by 'any *a*' is somewhat unclear. Russell (1903b, p. 57) also seems to say that the concept *any a* denotes a *variable conjunction* of terms. It is not completely clear how this can be squared by saying that the concept *any number* denotes 'one number, but not a particular one' (p. 91). It is difficult to see what can be meant by saying that some *one* term is a variable *conjunction* of terms. There is a further complication. In one of the quotes displayed and discussed above, Russell asks 'in which sense "any number" can be taken to be a definite object' (p. 91). Now in the *Principles of Mathematics*, Russell uses quotation marks in what from a contemporary perspective appears to be a queer manner: an expression of the form "*a*" refers to the *concept* expressed by the expression '*a*' rather than to the expression '*a*' (p. 53). But it is clear that the *concept* expressed by the expression 'any number' is not a number at all. The question that one would have expected Russell to ask at this point is how the *denotation* of the concept 'any number' relates to the specific numbers 1, 2, 3, ... Because of these interpretational difficulties, I surmise that Russell had not, at this point, worked out this particular aspect of his account in full detail.

Russell's considered view in 1903 is in any case a bit more complicated than it appears at first sight, for he recognises that *correlation* of values of variables presents a complication (p. 94):

Thus x is, in some sense, the object denoted by any term; *yet this can hardly be strictly maintained, for distinct variables may occur in a given proposition, yet the object denoted by* any term, *one would suppose, is unique. This, however, elicits a new point in the theory of denoting, namely that* any term *does not denote, properly speaking, an assemblage of terms, but denotes only one term, only not one particular definite term. Thus* any term *may denote different terms in different places. We may say: any term has some relation to any term; and this is quite a different proposition from: any term has some relation to itself. Thus variables have some kind of individuality ... A variable is not* any term *simply, but any term as entering into a propositional function.*

Moreover, in that same year, Russell (1903a, p. 334) recognised – due no doubt to the fact that the argument of the Russell paradox continued to exercise him – that type-distinctions have to be made in the range of variables:[2]

---

[2] A good discussion of this is given in Ito (2017, Section 4.4.2).

If we want a variable whose values are to be dependent variables containing x, we must have a new variable of a different kind.

Russell (1903b, p. 94) conceded that all this does not amount to a fully worked out theory of variables as they are used in mathematics. From our current vantage point, we take Russell to have given us a somewhat confusing and incomplete version of an *obsolete* conception of variable. But Russell's incomplete theory does contain most of the key ingredients of a theory of arbitrary objects and for that reason merits our attention. In the next sections, I will attempt to articulate this theory in some detail. I will draw on elements of Fine's theory and of Russell's theory. But the theory I propose differs from both of these.

3.3 Arbitrary Objects and Their Relations

Arbitrary objects of a given kind F are *abstract* entities that *can be in a state* that belongs to the *state space* associated with F. For instance, an arbitrary person is an abstract entity that can take any value taken from the class of all *specific* people, which is its associated state space. Objects that are not the kinds of things that can be in such states, such as rocks and vacuum cleaners, are called *specific objects*.

There is an intuitive distinction between *independent* arbitrary Fs and *dependent* arbitrary Fs: it is natural to say that an arbitrary object b (functionally) depends on an arbitrary object a if the value that b takes depends on the value that a takes. Suppose someone says the following:

Example 3.2.
    Consider an arbitrary human (a).
    Now consider his/her mother (b) ...

Then b is also an arbitrary human. But while a is the independent arbitrary human (it is introduced as such), b is a dependent arbitrary human (it is introduced as such).

In this way, *chains* of dependencies can be formed. We could, for instance, now go on to discuss the father of b. In fact, arbitrary objects can depend on more than one arbitrary object. For instance, if we have two arbitrary humans b_1 and b_2, then we can talk about the (unique) youngest nearest common ancestor (b_3) of b_1 and b_2. Then the value of b_3 depends *both* on the value of b_1 and on the value of b_2.

This talk of dependence is suggestive, but it is in my view also somewhat misleading. Dependence between arbitrary objects is ultimately a matter of correlation between states. To say that b in example 3.2 depends on a means *no more* than that in every situation in which a is in the state of being a specific person, b is in the state of being the mother of that person. This is because if we know, for every possible situation, which state a given arbitrary object is in, then we know all there is to know about the *nature* of the arbitrary object in question. (This arbitrary object may in addition be your favourite arbitrary object, for instance, but that is another matter.) In this sense, the notion of arbitrary object is a *thin* notion.

3.4 Identity and Comprehension

Frege taught us that if you want to construct a theory of classes, the first things to look for are a criterion of identity and a comprehension principle. For a theory of classes, the desired criterion of identity is the axiom of *extensionality*: a class x is identical with a class y if and only if x and y have the same members. As far as comprehension is concerned, the principle of *naive comprehension* immediately suggests itself: every condition on classes determines a class. Unfortunately, the naive comprehension principle is inconsistent, as Russell showed. Zermelo then provided a reasonable and consistent way to restrict comprehension, which went on to become one of the basic axioms of set theory, which is our favourite theory of classes. In sum, identity is easy, comprehension is more complicated.

The states that an arbitrary object can be in function as a *class* that is associated with the arbitrary object. So it seems legitimate to ask for a reasonable criterion of identity and for a reasonable comprehension principle for arbitrary objects.

Let us look at identity first. We may indeed ask with Frege how arbitrary objects differ from each other. The following *criterion of identity* for arbitrary objects goes some way to answering this question:

Thesis 3.3.
For any F, and any arbitrary Fs a and b:
$a = b \Leftrightarrow$ in every possible situation, the value taken by a is identical to the value taken by b.

I believe that Thesis 3.3 is correct. But it is not self-evident. (We will return to this principle in Section 8.4.2.)

Next, let us look at comprehension for arbitrary objects. The naive comprehension principle for arbitrary objects of a given kind K is this:

Thesis 3.4.
For any condition Φ that holds for every element of a non-empty set A of objects of kind K and only of those elements, there is an arbitrary object a that can be in the state of being any element of A and can be in no other state.

This thesis can be generalised in the straightforward way to relations:

Thesis 3.5.
For any n-place relation Φ that is satisfied by at least one n-tuple of objects of kind K, there are arbitrary objects $a_1, \ldots, a_n$ such that there is a situation in which a_1 is in state $x_1 \in K$ and $\ldots$ and a_n is in state $x_n \in K$ if and only if $\langle x_1, \ldots, x_n \rangle \in K^n$ satisfies Φ.

Russell's paradox teaches us that in its unrestricted form, Thesis 3.4 (and therefore also Thesis 3.5) is false. If you take the condition 'x is an arbitrary object that cannot be in the state of being that same variable object', and apply Thesis 3.4 to it, then you obtain a contradiction in the familiar way.

In other words, as it is with class theory, so it is with arbitrary object theory. The naive comprehension principle 3.4 must be restricted. One way of doing this is restricting the kinds K for which the comprehension principle holds to kinds consisting only of *specific* objects:

Thesis 3.6.
For any condition Φ that holds for every element of a non-empty set A of specific objects of kind K and only of those objects, there is an arbitrary object a that can be in the state of being any element of A and can be in no other state.

This restriction certainly blocks the paradoxes.

Ultimately, this restriction may well be judged to be too restrictive. Indeed, it seems merely a first step on the way to a *typed* approach to arbitrary object theory.[3] Arbitrary objects that are 'abstracted' from kinds of specific objects may be called *first-level* arbitrary objects. We could go on to 'abstract' second-level arbitrary objects from kinds consisting of specific objects and first-level arbitrary objects, and so on. I will develop this suggestion in some detail in Section 6.9. For the time being, however,

---

[3] Something like this is briefly hinted at in Fine (1985b, p. 30).

let us stick with the radical restriction of comprehension to conditions of specific objects.

Even Thesis 3.6 and its straightforward generalisation to relations are still widely regarded to be untenable. They are taken to be refuted by the following argument, which seems due to Berkeley (1710, Introduction, X).[4] I here formulate it for a particular kind of objects (the natural numbers), but it is clear that the argument is general:

Let b be the independent arbitrary natural number (or a independent arbitrary natural number, if you think there are more than one). The arbitrary natural number b is a natural number. Every natural number is either even or odd. So b is either even, or odd. But by Thesis 3.6, b is neither even, nor odd. Contradiction.

The conclusion that is typically drawn from arguments such as this is that the concept of arbitrary object is fundamentally confused. Rescher (1958, p. 117) puts it as follows:

To regard a 'random element' as an element or a 'random individual' as an individual is to commit what Whitehead terms the 'fallacy of misplaced concreteness' and involves what philosophers have come to call a category mistake.

In the history of philosophy, and in the history of British empiricism in particular, this argument was considered of great importance. Up to the time of Berkeley, the notion of arbitrary object was at times invoked by empiricists in an uncritical manner. In an infamous passage in the *Essay*, Locke, for instance, gave the following ontological account of the 'general triangle' (Locke 1960, Book IV, Chapter 7, Section 9):

For example, does it not require some pains and skill to form the general idea of a triangle (which is yet none of the most abstract, comprehensive, and difficult) for it must be neither equilateral, equicrural [isosceles], nor scalenon; but all and none of these at once. In effect, it is something imperfect, that cannot exist; an idea wherein some parts of several different and inconsistent ideas are put together.

Clearly this is sloppy metaphysics at best, and it was felt that Berkeley was right to reject Locke's concept of 'general ideas' and to advocate an alternative view. Hume, for instance, writes (Hume 1739, Book I, Part I, Section 7):

---

[4] It is clear from the passage that was quoted in Section 3.2 that Russell was keenly aware of this problem.

A great philosopher [Berkeley] has disputed the received opinion in this particular, and has asserted that all general ideas are nothing but particular ones annexed to a certain term, which gives them a more extensive signification, and makes them recall, upon occasion, other individuals which are similar to them. As I look upon this to be one of the greatest and most valuable discoveries that have been made of late years in the republic of letters, I shall here endeavour to confirm it by some arguments, which I hope, will put it beyond all doubt and controversy.

However, the naive metaphysician is not so easily defeated. The fundamental question is: *is* the arbitrary number b a natural number? I say that it is not, and this is (in my view) precisely where Berkeley's argument goes wrong. *The* natural numbers are the *specific* objects $0, 1, 2, 3, \ldots$ The domain of arbitrary natural numbers will be seen to, in a sense, contain the natural numbers as limiting cases (Section 4.1). The domain of arbitrary numbers can therefore be seen an *extension* of the natural numbers. So b is not a natural number. To believe that it is, is to be deceived by language. We will also see in Section 4.1 that elementary properties and relations of/on the natural numbers can be very naturally lifted ('pointwise') from specific numbers to arbitrary numbers. So arbitrary natural numbers to some extent *look like* natural numbers; but they aren't natural numbers.

3.5 State Spaces

We have seen in Section 3.3 that for every kind F, the notion of being an arbitrary F comes with a state space (or value range, in Fine's terminology).

Consider, within the context of the natural numbers, some arbitrary natural number a that is always in the state of being a (specific) natural number smaller than 2. Now consider some extension of the natural numbers, say, the countable ordinal numbers. Then there is no non-arbitrary way of *identifying* a with any arbitrary countable ordinal that only ever takes a value smaller than 2. The problem is that a is underspecified: it is not coordinated with values of arbitrary countable ordinals that can be in a state of being an infinite ordinal. This point is of course familiar from the discussion about mathematical structuralism (see Section 5.5.1), and it generalises. In this sense, the notion of being an arbitrary F is a notion of *type*.

We can take *all* specific objects as a state space and consider the 'arbitrary type' O over it. This is in a sense a *universal* type, except for the fact that we are barring higher-order arbitrariness here. The class of arbitrary countable ordinals is a sub-class of *O*, and so is the class of all arbitrary natural numbers. Within this 'universal' context, the arbitrary natural numbers in fact form a sub-class of the arbitrary countable ordinals. So from this point of view, any arbitrary natural number *is* some arbitrary countable ordinal. But ordinarily we work in local contexts, and then it makes no sense to identify elements of different types of arbitrary objects.

I have suggested earlier that because of Russell's paradox, it is hard to see how there could be a truly universal type, the value range of which consists of *all* entities, specific and arbitrary. Indeed, consider the arbitrary object *a*, which can be in the state of being any entity *x* if and only if *x* cannot be in the state of being *x*. Then *a* can be in the state of being *a* if and only if it cannot be in the state of being *a*.

3.6 Being and Being in a State

Frege (1960) thought that the idea of indefinite or arbitrary numbers rests on a confusion. He added to Berkeley's reservations about the concept in the following passage (p. 111):

The expression 'a variable assumes a value' is completely obscure. A variable is to be an indefinite number. Now how does an indefinite number set about assuming a number? For the value is obviously a number. Does, e.g., an indefinite man likewise assume a definite man? In other connexions, indeed, we say that an object assumes a property; here the number must play both parts; as an object it is called a variable or a variable magnitude, and as a property it is called a value. That is why people prefer the word 'magnitude' to the word 'number'; they have to deceive themselves about the fact that the variable magnitude and the value it is said to assume are essentially the same thing, that in this case we have not *got an object assuming different properties in succession, and that therefore there can be no question of variation.*

We already have the means to dispel some of these worries; but some of Frege's concerns in this quote cannot be fully addressed until later in this book.

First of all, an arbitrary natural number is not a natural number, as Russell already saw: an arbitrary number is an entity of a higher type than a specific natural number (see Section 3.2). But Russell (1903b, p. 91) raised a deeper question: do arbitrary numbers deserve to be called *numbers* at all? It is not clear whether a definitive list of necessary and sufficient criteria for being a number can be drawn up.[5] On the one hand, we will see that elementary arithmetical operations can naturally be defined for arbitrary natural numbers. On the other hand, the arbitrary natural numbers are not linearly ordered in any straightforward sense. This last issue will be looked into in some detail later in this book, in Section 9.7.1.

Second, Frege is right that there is an important analogy between the relation between a changing physical object on the one hand, and an arbitrary number on the other hand. A physical object can take on different shapes over time but it is something over and above its time-slices (or even the set or mereological sum of them). Consider the Lena river. This river can be in different states: it can be in a low state, it can be flooding, it can be muddy. But we also say:

The Lena river is very long.

This means that even though in every possible situation the Lena river is in a state, it will exist at this possible state as a river as well as the state that it is in, and the two are not the same thing. Similarly, a variable number is something over and above the states that it can be in. The dimensions involved in the two cases are different: time on the one hand, a modal dimension on the other hand. It would be wrong to say that an arbitrary object *changes* along its modal dimension, for, unlike time, the modal dimension does not come with an ordering of the states. Nonetheless, a question remains: what is the modality that is expressed by 'can' in this context?

One might think that since we are investigating the *metaphysical* nature of arbitrary objects, *metaphysical possibility* is operative here. But this is not the case. A clear indication that a different kind of modality is at play here results from reflecting on what the actual 'world' associated with the modality is. An arbitrary F *might have been* in the state of being this specific F and *might have been* in the state of being that specific F, but isn't *actually* in any of those. For instance, it makes no sense to ask who the person on the Clapham omnibus *actually* is. All we can (loosely) say is that it *could* be this or that specific person; it could be you, and it could be me.

You may be tempted to think that we are nonetheless dealing with ordinary metaphysical possibility here, but that it is just that arbitrary objects are not *actually* in any specific state. But this is not right. We may, for instance, consider possible worlds – ways our world might have been – in

---

[5] In correspondence with Weierstraß, Cantor argued that infinitesimals should not be called numbers; see Cantor (1887).

which there are unicorns. It does not make sense to ask whether in *any* such world the man on the Clapham omnibus is in the state of being my next door neighbour. From this I conclude that the modality expressed in 'arbitrary object *a* can be in state *x*' is a *sui generis* modality; I call this notion of possibility *afthairetic possibility*.

It must be conceded that even if the modality associated with arbitrary objects *were* metaphysical possibility, it would be better to speak of states rather than of possible worlds (Burgess 2009, p. 42):

The 'worlds' terminology, unlike the 'states' terminology, tends to carry the connotation that the possibilities being contemplated are maximally specific. Thus in some examples the only 'states' that would need to be considered might be those in which a certain coin comes up heads and those in which that same coin comes up tails, while to each of these states would correspond a vast infinity of 'worlds' differing in the conditions of coins other than the one in question as well as in an unlimited range of other respects. Since maximal specificity is often not needed, this difference of connotation is one of the reasons to prefer the 'states' to the 'worlds' terminology.

Another reason is that the high-flying terminology of 'worlds' tends to provoke philosophical puzzlement in ways the more pedestrian terminology of 'states' does not. For example, there is no obvious absurdity in the assumption that the whole system of possible states of the world exists, and that when we speak of merely possible states of the world, what is merely possible is not the existence of the state, but rather the world's being in it. By contrast, the corresponding assumption that a whole system of merely possible worlds exists, which seems to imply that a whole population of merely possible people exist (since worlds tend to be full of people), can easily be made to seem absurd. For what is to say that all these people are possible but not actual, except that there could have been such people but there aren't? But then if one says that there are such merely possible people, one is saying that there are things that there aren't.

Burgess does not have afthairetic possibility specifically in mind in this passage. But his observations certainly *also* hold for the notion of possibility that is associated with arbitrary objects, even though for them, the notion of an 'actual' state makes no sense.

All this shows how arbitrary objects differ from *fictional entities*. Fictional objects do not exist in the actual world, but it is metaphysically possible for them to exist; arbitrary objects are not in any state even though they can be in states. But being in a state is not the same as existing. And metaphysical possibility is not the same as afthairetic possibility.[6]

States in this sense are not completely unfamiliar to us. The spinning top can be in many different states. The fair coin, as opposed to the physical coin that you can hold in your hand (see the quoted passage above), can be in exactly two equiprobable states. The fair die can be in any of six states (that are all equiprobable). The 'can' here is afthairetic possibility: it makes no sense to say that our world could have been such that it contained unicorns and the fair die came up six.

Afthairetic possibility is often masked by metaphysical possibility that is nearby. A specific concrete die that has been tossed and came up five could have come up three. This is a statement about metaphysical possibility. But when we say of the fair die that it can be in the state of coming up three, we are using afthairetic possibility. What does it mean to be in a state? I said that the man on the Clapham omnibus could *be* this or that specific person. But, as we have seen, there *is* more to being in a state than simply to be. Barring higher-order arbitrariness for the moment, for an arbitrary object to be in a state is to take a *specific* value. The man on the Clapham omnibus is *necessarily* the man on the Clapham omnibus. But the man on the Clapham omnibus, being an arbitrary object, cannot be in the state of being the man on the Clapham omnibus.

Another argument for the same conclusion is the following. The view that the man on the Clapham omnibus can literally *be* some specific person is vulnerable to Evans' (1978) argument against ontological vagueness. Suppose, without loss of generality, that there is a situation in which the man on the Clapham omnibus is Robert Jones. Necessarily, the man on the Clapham omnibus is the man on the Clapham omnibus. But Robert Jones is not in all situations the man on the Clapham omnibus. So there is a property (necessarily being the man on the Clapham omnibus) that the man on the Clapham omnibus has but Robert Jones lacks. Therefore, by Leibniz's law of the indiscernibility of identicals, Robert Jones cannot be the man on the Clapham omnibus.

In sum, when I said that the man on the Clapham omnibus could be me, I was speaking loosely and strictly speaking incorrectly. But now that it is clear how to correct such loose talk, I will continue to speak incorrectly in this way from time to time.

---

[6] This notion of afthairetic possibility is compared with physical, logical, and mathematical possibility in Section 8.2.1.

An actual physical object is in no way 'more than' what it is in this or that possible state of the world. So you might ask how an arbitrary object exists over and above its modal states. We are very near to metaphysical bedrock here. All I can say is that it exists over and above its modal states in a way that is similar to the way in which a physical object exists over and above, but not independent from, its temporal states (time slices).

3.7 Hairdressers

A worked-out simple example may help to illustrate what all this amounts to. Bearing in mind what was said in the previous sections, let us see how talk about arbitrary hairdressers can be modelled.

Let K be the collection of specific, flesh-and-bone hairdressers, and let $|K| = p$. More specifically, let K be enumerated as $k_1, k_2, k_3 \ldots, k_p$. We want to describe the collection **A** of arbitrary hairdressers.

Let us say that *fully arbitrary* hairdresser σ is an arbitrary person who can be any hairdresser but cannot be anything but a hairdresser. So each such $\sigma \in$ **A** can be a woman and can be a man, for instance, but cannot be a chair.

Then if there are any fully arbitrary hairdressers at all, the state space Ω associated with the collection of arbitrary hairdressers must contain at least p states. There is no need identify states in a more fine-grained way, i.e. to take Ω to consist of more than p states. So we say that $|\Omega| = p$, and we can label states by means of natural numbers. This will then determine the structure of arbitrary hairdressers in the following manner.

Consider the matrix $M = \{1, 2, \ldots, p\} \times K$ in Figure 3.1. The horizontal dimension varies over the specific hairdressers; the vertical dimension varies over the states. Each entry $\langle a, b \rangle$ in M can be taken to stand for the object b in state a. Then **A** is the collection of functions $f : \{1, 2, \ldots, p\} \to K$: each such function σ specifies which specific object b the function σ is in state a (for all states a and specific objects b). In other words, arbitrary hairdressers are threads through M. The boxed entries in the matrix M form a (partial description of) a thread, i.e. an arbitrary hairdresser σ. This arbitrary hairdresser σ is in the state of being specific hairdresser k_2 at state 1, σ is k_1 at state 2, k_3 at state 3,..., k_p at state p.

We then see that $|\textbf{A}| = p^p$. Moreover, we see that there are $p!$ fully arbitrary hairdressers (the number of permutations of K), and that for every fully arbitrary hairdresser σ and every specific hairdresser k_i, there is *exactly one* state where σ is in the state of being k_i. Let us label arbitrary

$$\begin{array}{ccccc}
\langle k_1,1\rangle & \boxed{\langle k_2,1\rangle} & \langle k_3,1\rangle & \dots & \langle k_p,1\rangle \\
\boxed{\langle k_1,2\rangle} & \langle k_2,2\rangle & \langle k_3,2\rangle & \dots & \langle k_p,2\rangle \\
\langle k_1,3\rangle & \langle k_2,3\rangle & \boxed{\langle k_3,3\rangle} & \dots & \langle k_p,3\rangle \\
\vdots & \vdots & \vdots & \vdots & \vdots \\
\langle k_1,p\rangle & \langle k_2,p\rangle & \langle k_3,p\rangle & \dots & \boxed{\langle k_p,p\rangle}
\end{array}$$

Figure 3.1 An arbitrary hairdresser.

$$\begin{array}{ccccc}
\langle k_1,1\rangle & \boxed{\langle k_2,1\rangle} & \langle k_3,1\rangle & \dots & \langle k_p,1\rangle \\
\langle k_1,2\rangle & \boxed{\langle k_2,2\rangle} & \langle k_3,2\rangle & \dots & \langle k_p,2\rangle \\
\langle k_1,3\rangle & \boxed{\langle k_2,3\rangle} & \langle k_3,3\rangle & \dots & \langle k_p,3\rangle \\
\vdots & \vdots & \vdots & \vdots & \vdots \\
\langle k_1,p\rangle & \boxed{\langle k_2,p\rangle} & \langle k_3,p\rangle & \dots & \langle k_p,p\rangle
\end{array}$$

Figure 3.2 A specific hairdresser.

hairdressers such that for each specific hairdresser k_i there is exactly one state where σ is k_i as *diagonal hairdressers*. Then we can re-state the fact that we have just mentioned as follows: the collection of the fully arbitrary hairdressers is identical to the collection of diagonal hairdressers.

Specific hairdressers are canonically embedded in **A** as *constant functions*: we can identify each specific hairdresser $k_i \in K$ with the function $\sigma_i : \Omega \to k_i$. For instance, the specific hairdresser k_2 is canonically embedded in the set of arbitrary hairdressers as the boxed vertical column in Figure 3.2.

We could go on to introduce higher-order arbitrary hairdressers, but then, as we know from Section 3.4, we have to take care not to get entangled in the Paradox of the Barber as described in Russell (1919, Section 6).

3.8 Baby Steps

I now extrapolate from the example of arbitrary hairdressers and take some baby steps in the direction of a *general theory of arbitrary objects*. I merely seek to identify a few central concepts, and am not much concerned with exploring their properties.

3.8.1 Total arbitrary object spaces. Suppose we are given a kind F of specific objects, with $|F| = \kappa$. Then we want to associate an *arbitrary object space* with F in a systematic manner. We can do this by defining the *complete arbitrary object space associated with F* as

Definition 3.7.

$$\mathsf{A}(F) \equiv \{f : dom(f) = F \text{ and } ran(f) \subseteq F\}.$$

Of course we can take subsets of $\mathsf{A}(F)$ as 'smaller' or 'incomplete' total arbitrary object spaces with F. The domain of the functions plays the role of the *state space* $\Omega(F)$ that is associated with F.

There will then be a distinguished set $\mathsf{A}_{ct}(F) \subseteq \mathsf{A}(F)$ of constant functions:

Definition 3.8.

$$\mathsf{A}_{ct}(F) \equiv \{f \in \mathsf{A}(F) : \exists c \in F \forall s \in \Omega(F)(f(s) = c)\}.$$

The elements of $\mathsf{A}_{ct}(F)$ are the canonical representatives of the elements of F in $\mathsf{A}(F)$.

In this way, for every property of objects F in a domain of discourse $\mathbf{D}$, there is a complete arbitrary object space associated with F. So if we take every subset of $\mathbf{D}$ to determine a property, then a *collection of arbitrary object spaces* is associated with $\mathbf{D}$:

Definition 3.9.

$$\mathcal{A}(\mathbf{D}) \equiv \{\mathsf{A}(F) : F \subseteq \mathbf{D}\}.$$

Boolean operations can be defined in a natural way on the elements of $\mathcal{A}(\mathbf{D})$:

Definition 3.10.

1   $\mathsf{A}(F) \vee \mathsf{A}(G) \equiv \mathsf{A}(F \cup G)$;
2   $\mathsf{A}(F) \wedge \mathsf{A}(G) \equiv \mathsf{A}(F \cap G)$;
3   $\overline{\mathsf{A}(F)} \equiv \mathsf{A}(\mathbf{D} \backslash F)$.

So if we take $\mathcal{A}(\mathbf{D})$ to equipped with these operations (including the natural infinitary versions of the Boolean operations $\vee$ and $\wedge$), we have

Proposition 3.11. $\mathcal{A}(\mathbf{D})$ is a complete Boolean algebra.

$\mathcal{A}(\mathbf{D})$ is the most *liberal* collection of arbitrary object spaces associated with a domain $\mathbf{D}$. But you might want to be more restrictive about the properties that have arbitrary object spaces associated with them. For instance, for certain purposes you might want to restrict this collection to arbitrary object spaces that are associated with *natural kinds*. This will in general result in a proper subset of $\mathcal{A}(\mathbf{D})$.

There is a natural way of forming a new arbitrary object space from a given arbitrary object space by *restriction* to a sub-kind:

Definition 3.12. Suppose $G \subseteq F$.

$$\mathsf{A}(F)|G \equiv \mathsf{A}(F) \cap (G \times G)$$

Then we see that

Proposition 3.13.

$$\mathsf{A}(F)|G = \mathsf{A}(G).$$

3.8.2 Sub-spaces and partial arbitrary object spaces. The space $\mathcal{A}(\mathbf{D})$ treats the arbitrary Fs and the arbitrary Gs as un-coordinated with each other. But you may instead want to recognise *dependencies* between the arbitrary Fs and the arbitrary Gs. In general, if $F \subset G$, then you may be interested in the arbitrary Fs and the arbitrary Gs as structures in themselves; but you may also be interested in the arbitrary Fs as forming a *sub-structure* of the arbitrary Gs.

For instance, suppose we have a collection F of teachers and a sub-collection G of head teachers. Then we may want to take some arbitrary teacher and some arbitrary head teacher to take the same value in some state. To keep matters simple, suppose that $F = \{t_1, t_2, t_3\}$, and $G = \{t_1, t_2\}$. Then the space $\mathsf{A}(F)$ of arbitrary teachers consists of the threads through the matrix in Figure 3.3.

The space $\mathsf{A}_G(F)$ of arbitrary head teachers. is uniquely embedded in $\mathsf{A}(F)$. It can be seen as the (incomplete) *dependent sub-space* of $\mathsf{A}(F)$ that is defined as follows:

Definition 3.14. $\mathsf{A}_G(F) \equiv \{f \in \mathsf{A}(F) : \forall s \in \Omega(F)(f(s) \in G)\}.$

It is clear that there is a close connection between $\mathsf{A}_G(F)$ and $\mathsf{A}(F)|G$: the arbitrary object space $\mathsf{A}(F)|G$ can be obtained from $\mathsf{A}_G(F)$ by 'deleting' $|F| - |G|$ states from the state space of $\mathsf{A}_G(F)$.

$\langle t_1, 1 \rangle$	$\langle t_2, 1 \rangle$	$\langle t_3, 1 \rangle$
$\langle t_1, 2 \rangle$	$\langle t_2, 2 \rangle$	$\langle t_3, 2 \rangle$
$\langle t_1, 3 \rangle$	$\langle t_2, 3 \rangle$	$\langle t_3, 3 \rangle$

Figure 3.3 Total space of arbitrary teachers and head teachers.

Observe that the space $\mathbf{A}_G(F)$ does not contain a *diagonal head teacher*, i.e. an arbitrary head teacher that takes each specific head teacher *exactly once* as its value. Indeed, any fully arbitrary head teacher in $\mathbf{A}_G(F)$ takes one specific value once, and another specific value twice. (The boxed thread in Figure 3.3 is one example of such an arbitrary head teacher.)

If every state in $\Omega(F)$ is equally likely, then this implies that $\mathbf{A}_G(F)$ contains no arbitrary head teacher that is equally likely to be in the state of being t_1 as in the state of being t_2. This seems undesirable.

This consequence can be avoided by working in a slightly more general setting. We consider, instead of arbitrary objects that take a value at every state, arbitrary objects that may be *undefined* in some states. In other words, for every kind $F \subseteq \mathbf{D}$, instead of Definition 3.7, we define the *partial arbitrary object space* $\mathbf{A}^P(F)$ associated with F:

Definition 3.15.

$$\mathbf{A}^P(F) \equiv \{f : dom(f) \subseteq F \text{ and } ran(f) \subseteq F\}.$$

It is easy to see that $|\mathbf{A}^P(F)| = (\kappa + 1)^\kappa$.

To illustrate this, we return to our example of teachers and head teachers. If we denote absence of value as $\star$, then we can depict the *partial* arbitrary object space $\mathbf{A}^P(F)$ as the matrix in Figure 3.4.

The space of arbitrary head teachers can be taken to be embedded as an incomplete *partial* sub-space $\mathbf{A}^P_G(F)$ of $\mathbf{A}^P(F)$ in the same way as above. But we see that, in contrast to $\mathbf{A}_G(F)$, the sub-space $\mathbf{A}^P_G(F)$ contains arbitrary head teachers that take each specific value *exactly once*. The boxed thread in Figure 3.4 – let us call it σ – is an example of such a diagonal head teacher. And if each state in $\Omega(F)$ is equally likely to obtain, then the probability that σ is in the state of being t_1 is equal to the probability that σ is in the state of being t_2.

$\langle t_1, 1 \rangle$	$\boxed{\langle t_2, 1 \rangle}$	$\langle t_3, 1 \rangle$	$\langle \star, 1 \rangle$
$\boxed{\langle t_1, 2 \rangle}$	$\langle t_2, 2 \rangle$	$\langle t_3, 2 \rangle$	$\langle \star, 2 \rangle$
$\langle t_1, 3 \rangle$	$\langle t_2, 3 \rangle$	$\langle t_3, 3 \rangle$	$\boxed{\langle \star, 3 \rangle}$

Figure 3.4 Partial space of arbitrary teachers and head teachers.

3.9 Properties, Quantities

The values of (first-order) arbitrary objects are specific objects. Even for arbitrary objects that have the same value range, there can be states in which they take different values. This is a prima facie difference between arbitrary objects and *sets*: the value range of an arbitrary object (i.e. a set) does not capture all there is to say about that arbitrary object.

If you believe that there are arbitrary objects, then you may wonder whether there are also *arbitrary properties*. The value of an arbitrary property, if there were any, would be a specific property. For instance, one could perhaps imagine the property of being an arbitrary colour as being in the state of being red or as being in the state of being blue. As with arbitrary objects, the values of arbitrary properties would be *coordinated*. There would be states in which one arbitrary colour is in the state of being red, whilst another arbitrary colour is in the state of being blue. And, as before, specific colours are embedded in this system of arbitrary colours as constant functions.

Properties are exemplified by objects. So one way of developing this idea in more detail would be to model the state space of an arbitrary property as a collection of ordered pairs of objects and times. Arbitrary colours could then be seen as (partial) functions from states to pairs of objects and times. However, I know of no one who has developed a general theory of arbitrary properties, or even elements of it. Nevertheless, we will see in Chapter 10 that there is a relation between the notion of arbitrary property on the one hand, and the notion of random variable on the other hand.

An idea in the neigbourhood of arbitrary properties, however, has been entertained. Some properties exist as a magnitude or as a multitude: they are *quantities*.[7] Many quantities can be put on an ordinal scale, or even measured by real numbers.

For this reason, quantities can be modelled as functions from pairs of objects and time to *numbers*. This is, of course, what science does all the time: it is the concept of a *scientific variable*. One way of spelling out in detail how this can be done is found in Menger (1954, pp. 135–136).

In earlier times, physical quantities were often seen as *variable entities* that are coordinated with other quantities. By analogy, mathematical quantities were then also seen as variable entities. The following passage by de la Valée Poussin, similar in spirit to Russell's 1903 theory, is a typical illustra-

---

[7] I will not even try to sketch the complicated history of the notion of quantity here.

tion of how French analysts thought of mathematical 'variables' (Poussin 1903, p. 8):[8]

On appelle variable une quantité qui peut recevoir une infinité de valeurs différentes. Soit u une variable qui passe par une infinité de valeurs réelles, rationelles ou non,

$$u = u_1, u_2, \ldots, u_n, \ldots, u_{n+p}, \ldots$$

One variable quantity can be taken as a standard against which the variation of other mathematical quantities is measured. For instance, the ω-sequence of the natural numbers can be taken as a standard, and other mathematical quantities whose value range is a subset of the natural numbers can be regarded as functions on the natural numbers. In this way, functions on the *specific* natural numbers can be seen as variable (or arbitrary) numbers.

Next, functions on the *variable* numbers can then be regarded as co-variation relations between variable quantities. In Euler's (1755, p. 4) words:

If some quantities so depend on other quantities that if the latter are changed the former undergo change, then the former quantities are called functions of the latter. This denomination is of broadest nature and compromises every method by means of which one quantity could be determined by others. If, therefore, x denotes a variable quantity, then all quantities which depend on x in any way or are determined by it are called functions of it.

One crucial difference between mathematical functions and physical quantities is that mathematical quantities, regarded as functions, are often iterable, whereas physical quantities are not (Menger 1954, p. 137). This is because the domain and the range of mathematical functions can be the same kind of entities (the specific natural numbers, for instance), whereas for physical quantities regarded as scientific variables, this is never the case. This phenomenon is one of the reasons why mathematical functions are so useful in the physical sciences.

Nonetheless, from a conceptual point of view, in this movement from physical quantities to mathematical quantities, conceptual unclarity arises. Menger rightly complains concerning definitions such as Poussin's, that

---

[8] Translation: 'A quantity that can receive an infinity of values is called a variable. Let u be a variable that passes through an infinity of real values, be they rational or not,

$$u = u_1, u_2, \ldots, u_n, \ldots, u_{n+p}, \ldots$$'

what it means for a mathematical variable to 'pass through' variables is never made clear (Menger 1954, p. 134). In other words, whereas for physical quantities it is reasonable to say that they vary over time, it has not been made clear what the domain of variation of mathematical quantities is.

In the next chapter, I will argue that mathematical arbitrary objects are a sub-species of arbitrary objects as described in the present chapter. This means that, in accordance with what was said in Section 3.6, their dimension of variation is afthairetic possibility instead of time. Thus arbitrary mathematical objects are of a different nature than physical quantities as traditionally conceived. Since, as we have seen, we are also naturally led to introduce a temporal component in the state spaces of 'arbitrary properties', I will leave both of them (arbitrary properties, physical magnitudes) outside the scope of my investigations in the remainder of this book.

4 | Mathematical Objects as Arbitrary Objects

> It used to be necessary to warn against the notion of variable numbers, variable quantities, variable objects, and to explain that the variable is purely a notation, admitting only fixed numbers or other fixed objects as its values. This dissociation now seems to be generally understood, so I turn to others.
>
> *(Quine [1975, p. 155])*

Until now, the discussion of arbitrary objects has been rather abstract and metaphysical. The aim of this chapter is to continue on the path that was taken in Section 3.7 of the previous chapter, i.e. to make arbitrary object theory somewhat more tangible by focussing on applications. In this chapter, I will apply it to examples related to mathematics.

For as long as there have been theories of arbitrary objects, many of the paradigmatic examples of arbitrary objects have been drawn from number theory (arbitrary natural numbers, for instance) and geometry (arbitrary triangles, for instance). In this chapter, we take a closer look at some examples of arbitrary objects that are related to number theory. In particular, I investigate the properties of arbitrary natural numbers and the epistemological importance of arbitrary finite strings of strokes, which can play the role of natural numbers, as Hilbert taught us more than a century ago.

Along the way, I will discuss (and dismiss) more objections, mostly due to Frege, against arbitrary object theory in general.

4.1 Arbitrary Natural Numbers

Let us first apply arbitrary object theory to a paradigmatic test case: the natural numbers. And let us, for convenience, speak somewhat loosely, as I already did on a few occasions in the previous chapter.

4.1.1 Motivation. There are *specific* natural numbers: $0, 1, 2, 3, \ldots$ But beside specific natural numbers, there are also *arbitrary* natural numbers. As we have seen, an arbitrary number can be thought of as what a mathematician refers to when she says

'Let *a* be a natural number...',

and then goes on to reason about *a*.

There are many arbitrary natural numbers. For instance, it would make perfect sense for our mathematician, in the course of her argument, to add

'Now let *b* be some *other* arbitrary natural number...',

and go on reasoning about both *a* and *b*.

In general, an arbitrary natural number does not have any specific natural number as its *determinate* value. There is no determinate matter of fact, for instance, about whether the value of our mathematician's arbitrary number *a* is 23 – remember: this is loose talk.

There *can* be a determinate fact about whether an arbitrary number *x* is numerically identical with an arbitrary number *y*. Our mathematician was perfectly within her rights when she required the arbitrary numbers *a* and *b* to be non-identical. She might just have said, more clearly perhaps:

'Take any two arbitrary numbers *a* and *b* such that $a \neq b$, ...'

When an arbitrary natural number does not determinately have some given specific number as its value, there is a sense in which it *can* be the specific number in question. Thus we say that arbitrary numbers can be in different specific *states*. There is, however, no *actual* state in which the arbitrary number is. We might say, perhaps, is that an arbitrary number 'actually' is in a 'superposition' of states, but the notion of modality that is operative here (afthairetic possibility) does not come with any preferred (or 'actual') situations.

There are *degrees* of arbitrariness. If our mathematician says

'Let *a* be an arbitrary natural number larger than 10',

then she refers to a number that is less arbitrary than when she says

'Let *a* be an arbitrary natural number larger than 5'.

4.1.2 Modelling. The arbitrary natural numbers can be modelled in a way that is analogous to the way in which the arbitrary hairdressers were modelled in Section 3.7. We want there to be as many states as are needed to have fully arbitrary natural numbers, i.e. arbitrary natural numbers that *can* be every specific natural number. Moreover, we do not seem to have reasons to have more states. This means that $|\Omega| = \omega$: see Figure 4.1. Arbitrary natural numbers can then be seen as threads through the matrix $\omega \times \omega$, i.e. as functions from ω to ω. For instance, the boxed thread is the arbitrary natural number σ that is the number 3 in state 1, the number 2 in state 2,

$$\begin{array}{ccccc}
\langle 1,1 \rangle & \langle 2,1 \rangle & \boxed{\langle 3,1 \rangle} & \langle 4,1 \rangle & \dots \\
\langle 1,2 \rangle & \boxed{\langle 2,2 \rangle} & \langle 3,2 \rangle & \langle 4,2 \rangle & \dots \\
\langle 1,3 \rangle & \langle 2,3 \rangle & \boxed{\langle 3,3 \rangle} & \langle 4,3 \rangle & \dots \\
\langle 1,4 \rangle & \langle 2,4 \rangle & \langle 3,4 \rangle & \boxed{\langle 4,4 \rangle} & \dots \\
\vdots & \vdots & \vdots & \vdots & \vdots
\end{array}$$

Figure 4.1 Arbitrary natural numbers.

the number 3 in state 3, and the number 4 in state 4. It is easy to see that there then are 2^ω arbitrary natural numbers.

Let us call this 'space' of arbitrary natural numbers **N**. Of course **N** contains some excess structure, for the state space, while being of cardinality ω, is not 'naturally ordered' in any particular way.

We can straightforwardly transfer some terminology and concepts that were introduced in Section 3.7 to arbitrary natural numbers.

Specific natural numbers are closely related to *limiting cases* of arbitrary natural numbers. For many purposes they can be identified with arbitrary natural numbers that have an absolutely minimal degree of arbitrariness:

Definition 4.1. A *specific* number is a *constant* function from ω to ω.

Next, there are a few ways of understanding what it means to be *highly arbitrary*:

Definition 4.2. An arbitrary natural number is *fully arbitrary* if it can be any specific natural number whatsoever.

Definition 4.3. An arbitrary natural number σ is a *diagonal number* if for every specific number m, there is exactly one state in which σ is in the state of being m.

Let **F** be the collection of fully arbitrary numbers, and let **D** be the set of diagonal numbers.

There then are 2^ω diagonal numbers, and therefore also 2^ω fully arbitrary numbers. For elementary cardinality reasons, the fully arbitrary numbers and diagonal numbers do not have quite the same properties as their hairdressing cousins:

Proposition 4.4.

1 $D \subset F$.
2 There are 2^ω arbitrary numbers $\sigma \in F$ that take each specific value in infinitely many states.

We can generalise the concept of diagonal number somewhat:

Definition 4.5. For every cardinal number κ, an arbitrary number σ is said to be κ-generic if for every $m \in \mathbb{N}$, σ takes the specific value m in κ many states.

Then the diagonal numbers are the 1-generic arbitrary numbers. But there are also ω-generic numbers, for instance.

In the case of the hairdressers, it was intuitively clear that the diagonal hairdressers are all and only the fully arbitrary hairdressers. In the case of the arbitrary numbers, it is not immediately clear which arbitrary numbers should be considered to be 'maximally arbitrary'. It is not intuitively clear, for instance, whether diagonal numbers are more or less arbitrary than ω-generic arbitrary numbers.

From an abstract point of view, there is an obvious way to generalise from this way of modelling arbitrary arbitrary natural numbers to modelling higher-order arbitrary natural numbers. A second-order arbitrary number is an entity that can be in a state of being different arbitrary natural numbers, and so on. This gives rise to the following definition of a *hierarchy* of higher-order arbitrary natural numbers:

Definition 4.6.

(1) $A_0 = \mathbb{N}$;
(2) $A_{\beta+1} = A_\beta^{A_\beta}$;
(3) $A_\lambda = \bigcup_{\beta < \lambda} A_\beta$ for λ a limit ordinal.

4.1.3 Properties. Following up on Euler's observation in Section 3.9, we note that arbitrary natural numbers can *depend* on each other in interesting ways. These dependence relations are not always functional. The numbers a and b in Section 4.1.1 do not stand in a *functional dependence* relation, yet they do depend on each other. For instance, if a takes the value 101, then b does not take the value 101, and vice versa. The notion of dependence that is relevant here is described in another context by Väänänen and Grädel (2013, p. 401) as follows:

Two variables are independent if observing one does not restrict in any way what the value of the other is.

In our setting, we capture this notion as follows:

Definition 4.7. Arbitrary numbers a and b are *independent* of each other if for every specific value k that a can take, and every specific value l that b can take, there is a situation where *a* takes value *k* and *b* takes value *l*; otherwise *a* and *b* depend on each other.

An arbitrary number can be more or less *likely* to be in a given state. For instance, an arbitrary natural number between 25 and 50 is unlikely to be a power of 2 (the probability is $\frac{6}{26}$). For 'finite' probability problems of this kind, correct answers are straightforwardly computed using the ratio formula. For 'infinite' problems, matters are not so simple. In the context of classical probability theory (Kolmogorov probability), it is not even clear that the question what the probability is that a diagonal natural number is the number 23, for instance, has a well-defined meaning.

Clearly the notions of dependence and independence, and the notion of probability that can be associated with arbitrary objects, merit closer attention. We shall look at them in more detail in Chapters 9 and 10, respectively.

In general, there is a rich body of mathematical knowledge on which we can draw for finding interesting properties of the arbitrary natural numbers. The space **N** of arbitrary natural numbers is known as *Baire space* (ω^ω). This space is very familiar to mathematicians, and especially in topology and in descriptive set theory it has been intensively studied. Topological concepts appear rather naturally in the theory of arbitrary objects. For instance, Santambrogio (1987) has developed a topological approach to concepts of *degrees* of arbitrariness.

Baire space is equipped with the following natural *topology*:

Definition 4.8. The Baire space topology is generated from the countable sub-base that consists of all open sets of the form

$$[\tau] \equiv \{\sigma \in \omega^\omega \mid \forall n \in dom(\tau) : \sigma(n) = \tau(n)\} \qquad \text{with } \tau \in \omega^{<\omega}.$$

This gives rise to a notion of *continuity* of functions on the arbitrary natural numbers:

Definition 4.9. A function $f : \omega^\omega \to \omega^\omega$ is continuous if and only if $f^{-1}(A)$ is open whenever $A \subseteq \omega^\omega$ is open.

We then see that the space **N** of arbitrary numbers (i.e. Baire space) is not compact, that **N** × **N** is homeomorphic to **N**, etc.[1] Thus, I claim that Baire space is actually of philosophical importance.

---

[1] Some of the basic topological properties of Baire space are discussed in Truss (1997, Chapter 10).

An important concept in this context is that of *density* (in the topology):

Definition 4.10. A set $A \subseteq \omega^\omega$ is *dense* if and only if for every non-empty open set $B \subseteq \omega^\omega : A \cap B \neq \emptyset$.

It is then natural to consider countable dense sets, and to consider dense open sets of **N**. For example, the set of all $f \in \omega^\omega$ for which there is an n such that

$$\langle f(n), f(n+1), \ldots, f(n+9) \rangle$$

is the sequence of the first 10 digits in the decimal expansion of the real number π, is both dense and open. And therefore the set of all $f \in \omega^\omega$ such that for every $n > 0$ there is an m such that

$$\langle f(m), f(m+1), \ldots, f(m+n-1) \rangle$$

is the sequence of the first n digits in the decimal expansion of π is a *co-meagre set*, where a set is co-meagre if it contains a countable intersection of dense open sets (Truss 1997, p. 258). The set of ω-generic numbers, for instance, is also a co-meagre set. The concept of co-meagreness is theoretically significant because any co-meagre set has a prima facie claim to being a *typical* set.

4.2 More Fregean Worries

Frege has given the most incisive critique of arbitrary object theory. We saw in Section 3.6 how Frege raised the question in what sense arbitrary numbers can be said to be *numbers* – and we have not fully answered this question yet – as well as the question what it means for an arbitrary object to be in a state. Let me now try to answer further concerns that Frege raised in connection with arbitrary numbers.

In his article *What is a function?* (1904), Frege (1979, p. 160) asks further pointed questions about arbitrary or 'indeterminate' numbers:[2]

[The mathematician Czuber's account] gives rise to a host of questions. The author obviously distinguishes two classes of numbers: the determinate and the indeterminate. We may then ask, say, to which of these classes the primes belong, or whether maybe some primes are determinate numbers and others

[2] Frege's arguments against variable numbers are critically discussed in Santambrogio (1992).

indeterminate. We may ask further whether in the case of indeterminate numbers we must distinguish between the rational and the irrational, or whether this distinction can only be applied to determinate numbers. How many indeterminate numbers are there? How are they distinguished from one another? Can you add two indeterminate numbers, and if so, how? How do you find the number that is to be regarded as their sum? The same questions arise for adding a determinate number to an indeterminate one. To which class does the sum belong? Or maybe it belongs to a third?

These are valid questions: they need to be answered. We have seen in Section 2.4 that to the extent that we do not have the wherewithal to answer at least most of these questions, the whole enterprise of enquiring into the nature of arbitrary numbers is undermined. But we do have answers to Frege's questions:

1 The prime numbers are not arbitrary numbers: they are specific natural numbers. It may be that a concept of prime number can be generalised to the space of arbitrary natural numbers, just as it can be generalised to various algebraic structures, but that is another matter.

2 Being rational or irrational is a concept governing the specific real numbers. It may be that it can be generalised to the space of the arbitrary real numbers, but that is another matter.

3 In Section 4.1 it was argued that there are 2^{ω} arbitrary natural numbers. In Chapter 6 we shall see that if you hold a particular structuralist position about the natural number structure, then you will disagree with this argument and instead hold that there are $2^{2^{\omega}}$ arbitrary natural numbers.

4 In Section 3.4 I proposed an identity criterion for arbitrary objects according to which an arbitrary number a is identical with an arbitrary number b if and only if in each situation, a is in the same state as the state that b is in, or, in other words, if and only if a and b take the same specific value in each situation.

5 A natural notion of sum can be defined for arbitrary numbers in a pointwise manner. If a and b are arbitrary numbers, then their sum $a + b$ is the arbitrary number such that in each state where a is the specific number m and b is the specific number n, $a + b$ is the specific number $n + m$.

6 Given that a specific number can be seen as a limiting case of an arbitrary number (see Section 4.1), the sum of a specific natural number and a non-specific natural number is easily seen to be a non-specific arbitrary number.

4.3 Quasi-concrete Objects

Some objects that are ordinarily thought of as specific objects are really arbitrary objects. A symbol (a letter, for instance) is an arbitrary entity that can be this or that particular inscription (e.g. a chalk trail on a blackboard, a trace of ink on a piece of paper). Symbols can be combined into complex expressions by means of concatenation. So complex expressions are arbitrary objects that are correlated with the symbols out of which they are built.

Hilbert saw that arithmetic can be taken to be about a *systems of expressions*. For instance, it can be taken to be about finite strings of strokes. It would be unreasonable to take arithmetic to be exclusively about one preferred system of expressions. Why would it be about stroke notation, say, rather than about the system of roman numerals? So one might say that on a Hilbertian view, the natural number system is a *second-order* arbitrary entity that can be in the state of being the stroke notation system and can also be in the state of being the system of roman numerals, where a finite string of strokes or a roman numeral is itself a (first-order) arbitrary object.

Along similar lines, finite graphs can be taken to be arbitrary objects. Points and lines are arbitrary objects that can be in the state of being physical inscriptions (marks on a sheet of paper). Points and lines connecting them can be combined into diagrams that we recognise as finite undirected graphs. But the diagrams do not have to be built from straight lines and points. Alternatively, they can be constructed from cords and coffee mugs. So finite graphs are again *second-order* arbitrary entities.

We can view algorithms as arbitrary entities that can be in the state of being this computer programme or that computer programme, where a computer programme is in turn an arbitrary object that can be in the state of being written down on a piece of paper, or in the state of being implemented on a laptop computer.

Expressions, graphs, and perhaps even computer programmes, are examples of of the kind of abstract objects that Parsons (2008, Chapter 1) calls *quasi-concrete objects:*[3] they are distinguished by having an intrinsic relation to the concrete. There are no letters, for instance, that do not have concrete realisations. Their interest lies in the fact that they so to speak mediate between the concrete world and the abstract world. Since a significant part of our epistemic grasp of the abstract world of mathematical entities

---

[3] Linnebo (2018) calls them *thin* objects.

is grounded in our epistemic grasp of the concrete reality that surrounds us, quasi-concrete objects are of obvious importance to the philosophy of mathematics.

We might even go further and ask whether perhaps many ordinary physical objects should also be taken to be arbitrary objects. Take the raspberry bush in the allotment. Might it be an arbitrary object that can be in states that are objects that have sharp boundaries? This would be an ontological (rather than the usual semantic) interpretation of the supervaluation theory of vagueness as described in Fine (1975).[4] Pursuing this question further would take us too far afield.

4.4 Knowledge of Arbitrary Strings

Arbitrary objects are *abstract*. This raises the question how we as concrete, biological entities have knowledge of them (Benacerraf 1973). Frege saw this as a challenge that simply cannot be met (Frege 1984, p. 195):

do we not use 'x', 'y', 'z' to designate variable numbers? This way of speaking is certainly employed; but these letters are not proper names of variable numbers in the way that '2' and '3' are proper names of constant numbers; for the numbers '2' and '3' differ in a specified way, but what is the difference between the variables that are said to be designated by 'x' and 'y'? We cannot say. We cannot specify what properties x has and what differing properties y has. If we associate anything with these letters at all, it is the same vague image for both of them.

In the case of abstract objects that play an essential role in well-confirmed empirical science, appeal is often made made, in Quinean vein, to abductive reasoning (inference to the best explanation). Real numbers may be *postulated* to exist, perhaps, with as much right as unobservable entities in physics (such as neutrinos) are postulated. But this strategy seems not to be available for arbitrary objects: they do not appear to play an essential role in well-confirmed empirical science – not yet, anyway.

A better way of acquiring knowledge of arbitrary objects is to come to understand the *nature* of arbitrary objects better. Fine has taken steps in this direction, as we shall see in Chapter 7, and I have tried to contribute to this effort in the preceding sections. But there is also a role for intuition

---

[4] Fine mentions a possible connection between arbitrary object theory and supervaluational truth in Fine (1985b, p. 45).

in our knowledge about arbitrary objects. Some arbitrary objects can be *imagined*, and properties of the imagined objects can be *seen* to hold.

Parsons (2008) investigates the role of intuition in arithmetical knowledge. A special role is reserved for *quasi-concrete* objects: abstract objects of which it is essential that they have concrete exemplifications (see Section 4.3). Finite strings of strokes are such quasi-concrete objects. Stroke notation is a system that is isomorphic to the natural number structure, and Parsons claims that we have *intuitive* knowledge about quasi-concrete objects such as words and strings of strokes (p. 154):

> *Typically, the hearer of an utterance has a more explicit conception of what was uttered (e.g. what words) than he has of an objective identification of the event of the utterance. I believe that the same is true of some other kinds of universals, such as sense-qualities and shapes. Indeed, in all these cases it seems not to violate ordinary language to talk of perception of the universal as an object, where an instance of it is present. This is not just an overblown way of talking of perceiving an instance as an instance (e.g. seeing something red as being red), because the identification of the universal can be firmer and more explicit than the identification of the object that is an instance of it.*

On the face of it, it seems that arbitrary objects play an essential role in Parsons' (2008, pp. 173–174) account:

> *If we imagine any string of strokes, it is immediately apparent that a new stroke can be added. One might imagine the string as a* Gestalt, *present at once: Then, since it is a figure with a surrounding ground, there is space for an additional stroke. However, this leaves out an important aspect of the matter, since the imagination of an* arbitrary *string in this way will have to leave inexplicit its articulation into single strokes [...]*
>
> *[This] way of imagining an arbitrary string involves imagining a string of strokes without imagining its internal structure clearly enough to imagine a string of n strokes for some specific n. Such an imagining is common enough. I might for example imagine the crowd at a baseball game, without imagining a crowd consisting of 34,793 spectators.*

However, Parsons hastens to disabuse his readers of the impression that in imagination of an arbitrary string of strokes, there is any kind of indeterminacy in the string imagined (p. 174):

> *That there is no n such that I imagine it to have n strokes does not imply that I imagine it as not having a definite number of strokes. We might speak of the*

imagination as vague, but that can be misleading since it suggests that what I imagine *is indeterminate in some way.*

But then Parsons owes us an account of the relation between the imagination and the thing imagined. He does not give us one. If there is imagining an arbitrary string without arbitrary strings, then imagining must be imagining of a specific string. But which specific string is imagined in imagining an arbitrary string? Why one specific string (say, the string consisting of 34,793 strokes) rather than another? Moreover, the arbitrary imagination of a specific string (whatever that would turn out to be) would then have to give us *general* knowledge of the successor axiom for strings of strokes. I do not see how that is possible.

Santambrogio (1992) argues along similar lines, not in the context of Parsons' philosophy of mathematics, but in the context of a discussion of Tait's (1981) finitism as the latter expresses it.[5] Despite Tait's protestations to the contrary (p. 528), Santambrogio holds that Tait's appeal to arbitrary objects in his epistemological justification of finitist mathematics is unavoidable. He summarises his view as follows (Santambrogio 1992, p. 153):

The picture of arbitrariness resulting from all this is then the following. What turns an ordinary individual into an arbitrary one, thus justifying the generalization, is our seeing that it possesses a certain form and that it is in virtue of that form that it has the property we are interested in. As soon as we see that form as a reason for having the property in question, we thereby also know that any (other) individual with that form has the same property; ... This is just part of our seeing the form as a reason for having a property. There is no need to grasp any totality of objects in order to understand this, and the finitist can understand it as well as anybody else.

The more straightforward account of imagining an arbitrary string of strokes is not to reinterpret its content, but literally to take it to be the imagination *of* an arbitrary string. Moreover, the cognitive distance between grasping that an arbitrary object has a property and that all its values do, is minimal: if we see in imagination that an arbitrary string of strokes has a successor, then we see that all specific strings of strokes have a successor.[6]

In a similar way, we can see intuitively that addition on strings of strokes is a total operation, and we can verify its elementary properties. Suppose

---

[5] Jeshion (2014) and Incurvati (2015) also critically discuss Tait's and Parsons' attempts to justify the axiom that every string of strokes has an immediate successor; they do not in their discussion refer to Santambrogio (1992).

[6] Fine would say that we can conclude this by an application of the principle of *generic attribution*: see Section 7.3.3.

you have an arbitrary string a and an arbitrary string b in your imagination. Then you can move the string b into juxtaposition with a, i.e. you can *concatenate* them in your imagination. The result is the arbitrary string $a + b$. The recursive axioms governing addition on strings of strokes can be verified as follows. The verification of $a + 0 = a$ is of course trivial. Now consider the arbitrary strings a and b in your imagination. By appending a stroke to b (in your imagination), you obtain the string $b + 1$. Presently you want to verify that $a + (b + 1) = (a + b) + 1$. First decompose $b + 1$ into b and the one-stroke string by moving the last stroke of $b + 1$ away from b. Then concatenate a and b as before, yielding the string $a + b$. To conclude, concatenate $a + b$ with the remaining one-stroke string.

Thus imagination gives us intuitive general knowledge of arbitrary strings of strokes. Basically, it appears that *weak theories of concatenation* are intuitively verifiable in this way. For definiteness, consider the concatenation theory QT^+ proposed by Damjanovic (2017). Let a and b be two letters, and let $*$ express concatenation of strings. Then the axioms of QT^+ are (p. 384):

QT1 $x * (y * z) = (x * y) * z$;

QT2 $\neg(x * y = a) \wedge \neg(x * y = b)$;

QT3 $(x * a = y * a \rightarrow x = y) \wedge (x * b = y * b \rightarrow x = y) \wedge (a * x = a * y \rightarrow x = y) \wedge (b * x = b * y \rightarrow x = y)$;

QT4 $\neg(a * x = b * y) \wedge \neg(x * a = y * b)$;

QT5 $x = a \vee x = b \vee (\exists y(a * y = x \vee b * y = x) \wedge \exists z(z * a = x \vee z * b = x))$.

I claim that intuitive knowledge of the axioms of QT^+ can be obtained along the lines sketched above. (The reader may want to verify this.) There is nothing special about the particular concatenation theory QT^+ in this respect. We might as well have focused on Grzegorczyk's concatenation theory (as described in Švejdar 2009, p. 89), for instance. A story along these lines might also be told about other mathematical entities, such as finite graphs.

It is not clear how intuitive knowledge of a theory such as QT^+ can be obtained without appealing to arbitrary strings of strokes. At any rate, despite his claim to the contrary, Parsons' account of our intuitive general knowledge of strings of strokes makes essential use of arbitrary strings of strokes.

Concatenation theories such as QT^+ are of course mathematically very weak. Nonetheless, such theories can *simulate* a significant amount of mathematics, in the sense that they can *relatively interpret* mathematical theories with substantial content. First, it can be shown that weak concatenation theories interpret Robinson's theory of arithmetic, the system Q (Visser 2009; Švejdar 2009). Furthermore, it is known that Q interprets the system $I\Sigma_0$, i.e. the system that results from adding induction for quantifier-free formulae to Q (Ferreira and Ferreira 2013, Section 3). This implies that weak concatenation theory can interpret the elementary theory of Real Closed Fields, Euclidean geometry, and theories of Analysis that formalise elementary properties of real numbers and continuous functions (Damjanovic 2017, p. 381). Moreover, it can be shown that elementary concatenation theory interprets quantifier-free finitary set theory (and vice versa) (Damjanovic 2017). The theory $I\Sigma_1$ can of course *not* be interpreted in weak concatenation theories, for it proves the consistency of Q.

There is a close connection between weak concatenation theory, on the one hand, and what is called *strict predicativist arithmetic*, on the other hand. Nelson (1986) argued that first-order Peano Arithmetic, and even Primitive Recursive Arithmetic, codify an *impredicative* concept of natural number. For Peano Arithmetic, this is evident from the fact that arbitrary nestings of quantifiers are allowed in fomulae that instantiate the induction scheme: this betrays that the natural numbers are assumed to form a completed collection. For Primitive Recursive Arithmetic, this is shown by the fact that exponentiation, for instance, is assumed to be a total operation on the natural numbers, even though Σ_1-induction, and thus the assumption that the natural numbers form a completed whole, is required to *prove* that exponentiation is total. For this reason, Nelson believes that even Primitive Recursive Arithmetic exceeds the bounds of finitary arithmetic. This is a significant claim, for Tait (1981) has famously argued that Primitive Recursive Arithmetic coincides with finitary arithmetic. One candidate formalisation of the amount of arithmetic that can be known on strict predicativist grounds is given by Buss' system S_2^1 of bounded arithmetic. But S_2^1 can be interpreted in Q, and therefore in weak concatenation theory (Ferreira and Ferreira 2013, Section 4). Thus there may be a way in which intuitive sense can be made of finitary arithmetic on the basis of our intuitive knowledge of arbitrary string.

Simulating finitary arithmetic by means of a relative interpretation is one thing; verifying the *truth* of finitary arithmetic is another. Indeed, it is not clear to me whether the verification of the totality and of the elementary properties of string multiplication can be carried out in the imagination along the lines sketched at the beginning of this section. The same holds for quantifier-free induction on strings of strokes. It would seem, for instance, that in order to see that induction holds for strings of strokes, the concept of *iteration* is needed: a string of strokes would need to be *defined* as whatever can be obtained from the empty string by iteration of appending the one-stroke string. And, in any case, it is not true that the number three, for instance, *is* a string of strokes. So it is not clear that the natural numbers themselves can be imagined.

It is on grounds such as these that Parsons (2008, Chapter 6, Section 37) argues, convincingly in my view, that we cannot have intuitive knowledge of the natural numbers themselves. So the role of intuition even in arithmetical knowledge, is after all limited. If Parsons is right, then the number 8, for instance, cannot be imagined; but we can think of the number 8. Similarly, many arbitrary objects can be *thought of*. We will give examples in later chapters. We will also see that there are (many) arbitrary objects *a* such that we cannot, even in principle, think of *a*. But we are still able to quantify over them.

At this point, it may be worth stepping back and making the following observation. We started this section on the epistemology of arbitrary objects in the expected way: by considering arbitrary objects as a *challenge* for epistemology. But we have presently arrived at a way in which the theory of arbitrary objects can *assist* us in shedding light on epistemological problems.

4.5 Knowledge by Acquaintance and Knowledge by Description

De re knowledge of an object requires us to be *acquainted* with the object in question (Russell 1910). We are acquainted with concrete objects by standing in particular sorts of causal relations with them. The way in which we become acquainted with *abstract* objects is by their presence in our imagination. This generates intuitive *de re* knowledge of abstract objects.

We have seen that an arbitrary string of strokes can be imagined. We can even relate an arbitrary string in our imagination to other strings, for instance by appending a stroke to it. This is sufficient to yield *de re* knowledge of arbitrary strings. Nonetheless, Frege (1960, p. 109) is right that the discriminating power of our imagination is very restricted for arbitrary strings of strokes. He rightly observes that 'we cannot conceive of each variable [object] in its individual being'. Indeed, we will see that any 'variable' string of strokes is *absolutely undefinable*. Frege (1960, p. 110) goes too far when he concludes from this that 'there are thus no indefinite numbers' – after all, there surely are specific real numbers that we cannot conceive of in their individual being. But it does imply that our *de re* knowledge of arbitrary strings of strokes is limited.

We have seen in the previous section that it is harder to see how a natural number can be imagined: natural numbers live at a higher level of abstraction. So it is not clear that *de re* knowledge of natural numbers is possible in the strict sense of the word.

We can *define* each specific natural number. Indeed, more is true: we have *canonical notation systems* for natural numbers. In a *weaker sense* of 'knowing of', to know *of* a specific natural number *n* that it has the property φ amounts to knowing *that* $\varphi(\underline{n})$, where $\underline{n}$ is a canonical representation of *n* (the arabic numeral representation of *n*, for instance) (Horsten 2005). So in this weaker sense of the word, *de re* knowledge of specific natural numbers is possible. But since, as Frege observed, *arbitrary* natural numbers are not nameable at all,[7] we cannot have *de re* knowledge of them even in this weaker sense.

We can quantify over everything: I just did. So we can quantify over arbitrary objects. More in particular, we can quantify over arbitrary natural numbers and other arbitrary mathematical objects. By developing an *understanding* of the nature of arbitrary objects such as arbitrary natural numbers, we acquire *de dicto* knowledge about them. Thus we come to know general facts about their properties in relations. From an epistemological point of view, the situation we find ourselves in is not essentially different from the way in which we acquire knowledge about mathematical objects generally, except for the fact that most arbitrary natural numbers are highly undefinable.

4.6 Theory and Applications

The nature of arbitrary objects merits philosophical investigation in the first place because it is a fascinating and rich metaphysical subject in its own right. Moreover, the nature of arbitrary objects is at present poorly understood. As I said in the Introduction, I fear that we have rushed to applications too quickly.

The paradigmatic example of arbitrary objects remains that of arbitrary *numbers*, and the simplest kind of numbers are natural numbers. For this reason, natural numbers are given special attention in this book.

Having good mathematical models of what you are investigating is important in metaphysics. Set theory is the framework that is most suitable for developing such models. One of the aims is to develop set theoretic

---

[7] See Section 9.3.

models for arbitrary objects that are so faithful to the structure of them that they form a reliable guide to principles governing arbitrary objects themselves.

Fine has elaborated in much detail how we *reason* with arbitrary objects (Fine 1985a, 1985b). The incompleteness of arbitrary objects makes them well placed for use in quantificational logical reasoning. However, the last word has not yet been said about reasoning with arbitrary objects. Fine himself noticed that the central relation of dependence in arbitrary object theory points to a connection with *Skolem functions* (Fine 1985b, p. 46). From this observation it is only a small step to suggest that there therefore must also be a connection with *Dependence Logic*, as studied in Väänänen (2007), for instance. First steps in exploring this connection are taken in San Gines (2014), but there are many open questions here. Incidentally, in this connection it should also be mentioned that Fine's theory has been connected with Hilbert's *epsilon-calculus*: see Meyer Viol (1995, Chapter 2, esp. Sections 2.5–2.7).

Sometimes one part of metaphysics can be deeply connected to a seemingly quite *different* part of metaphysics: I have already mentioned (in Section 2.4) the relation between essence and modality in this context. Suggestions for relating the theory of arbitrary objects to other parts of metaphysics were made by Fine: they are mostly related to philosophy of language and to the philosophy and the foundations of mathematics. Indeed, Fine was from the beginning interested in *applications* of the theory of arbitrary objects (Fine 1983, pp. 73–77; 1985b, pp. 44–47), and he has worked some of them out in considerable detail.

Fine has suggested that the theory of arbitrary objects can be of use in *natural language semantics*. Some of this has been elaborated in work by other authors. King (1991) has followed up a suggestion by Fine (1983, p. 76) and worked on the relation between counterfactuals and arbitrary objects; the computer scientist Stuart Shapiro (2004) – not to be confused with the philosopher of mathematics Stewart Shapiro! – has used Fine's theory of arbitrary objects in the theory of anaphora.

Fine (2017a) has recently used elements of the theory of arbitrary objects to construct a theory of *form* of formulas. Fine (1983, p. 76) sees this as a prelude to a general theory of Form in the sense that philosophers like Plato or Locke saw it. Indeed, it might be fruitful to look at certain theories in the history of metaphysics through the lens of arbitrary object theory. In the previous chapter (Section 3.2) we briefly did so in our discussion of Russell's theory of variables in his *Principles of Mathematics*.

It has been suggested that arbitrary object theory can be used in modelling elementary (classical) Calculus (Fine 1983, pp. 74–75; Kripke 1992, p. 73), and infinitesimals (Fine 1985b, p. 43). As far as I know, these applications remain to be developed in detail. The theory of arbitrary objects might also provide a perspective on the interpretation of generic sets in the technique *forcing* in set theory (Fine 1985b, pp. 45–46). This is another connection that remains to be worked out.

As Kripke (1992, p. 72) remarked , the theory of arbitrary objects might be used to shed light on the notion of random variable in probability theory. And Fine himself has applied the theory of arbitrary objects to Cantor's theory of cardinal number and Dedekind's account of ordinal number (Fine 1998), and has outlined how the theory of arbitrary objects might be connected to structuralism in the philosophy of mathematics (Fine 1998, Section VI).

All this suggests that the concept of arbitrary object may be a theoretically fruitful notion. Nonetheless, in this book I will discuss only *two* of these potential applications. First, I will be much concerned with *structures* of arbitrary mathematical objects of certain kinds (natural numbers, countable graphs, etcetera). In this context I will also be much concerned with the relation between the theory of arbitrary objects and forms of non-eliminative structuralism in the philosophy of mathematics. In particular, Fine's outline of a theory of mathematical structure will be discussed in detail in Chapter 7. Second, in Chapter 10 I follow a lead of Kripke and explore to what extent arbitrary object theory can give a metaphysical account of random variables.

In sum, most of the potential applications of the theory of arbitrary objects that were identified by Fine will *not* be explored in this book; I regard some of these questions as premature at this stage of the investigation. Instead, I aim at laying the *groundwork* for exploring such potential applications by concentrating on basic metaphysical and logical aspects of arbitrary object theory in general. Nonetheless, at the end of this book, in Chapter 11, I briefly comment on what I regard as potentially promising avenues for future research that I am unable to pursue in this monograph.

5 | Structure in Mathematics

> [The argument] concludes ... that numbers couldn't be sets at all – and ... the same argument could be extended to any other 'objects' (not themselves obviously numbers) that we might choose to put in the place of sets ... The parenthetical reservation is necessary period.[1]
>
> *(Benacerraf [1996, pp. 22 and 51 n9])*

We are now changing gears and turning to the connection between arbitrary object theory and the notion of mathematical structuralism, which will occupy us for some time. The present chapter is of a transitional nature. Arbitrary object theory is put aside for a while, and we concentrate on forms of eliminative and non-eliminative structuralism in the philosophy of mathematics. The reason for doing so is that in the next chapters, when we will connect the notion of mathematical structure to arbitrary object theory, we will rely heavily on certain central concepts that have emerged in the literature on mathematical structuralism and that are essential ingredients in a good theory of the nature of mathematical structures.

5.1 Form and Structure

The concept of structure today arises in many contexts: one speaks of the structure of a novel, the structure of a molecule, the structure of a system of government. There may be different concepts of structure in different contexts, but they seem to be related.

The origin of our concept of structure lies in ancient Greece. Plato held that *Ideas* or *Forms* are abstract entities. They are the only objects of knowledge. The sensible material world around us consists of imperfect images of the Forms, and cannot really be known. Our sense perceptions are then mere images of images. The Forms are taken by Plato to be ontologically prior to the sensible world. The sensible world depends for its existence on the world of Forms, but not vice versa.

---

[1] "As George Boolos was quick to point out very long ago ('obviously' was, I believe, the technical term he employed)."

Because the sensible world consists of imperfect images of the Forms, empirical observation is not a suitable way of acquiring knowledge of the Forms. Plato thought instead that knowledge of the Forms is somehow innate in humans.

Aristotle agreed with Plato that knowledge is knowledge of Forms. But he disagreed with Plato's thesis on the ontological priority of the world of Forms over the sensible material world. Instead, he advanced a thesis of *co-dependence* of Form and matter (*hylemorphism*). Aristotle held that Forms can only exist incorporated in sensible objects: pure Forms cannot exist. This allowed Aristotle to be more optimistic about the possibility to understand the sensible world: understanding Forms is at the same time understanding the sensible world.

After the decline of Aristotelian philosophy, the debate of the mode of existence of the Forms or Ideas was transformed. The question whether Ideas exist independently of the human mind now replaced the question whether Forms can exist independently of the sensible world. To this new question, the empiricists said no, the rationalists said yes.

Sometime around 1900, the concept of Form or Idea metamorphosed into our notion of *structure*. The platonist idea that true knowledge is knowledge of Forms transformed into the thesis that scientific knowledge is knowledge of structure.

In contemporary philosophy, the notion of structure plays a central role in theories of our knowledge of the world in a way that is similar to the role of the notion of Form in the debate between Plato and Aristotle.

Some scientists and philosophers of science claim that science aims at revealing the structure of reality (Worrall 1989). Poincaré, for instance, claimed that the physical sciences give us knowledge of the *structure* of reality and nothing besides (Poincaré 1905, pp. 160–162). Some contemporary metaphysicians claim that structure is the fundamental constituent of reality. Sider is one of them. Sider (2011, p. vii) begins one of his books as follows:

The central theme of this book is: realism about structure. The world has a distinguished structure, a privileged description. For a representation to be fully successful, truth is not enough; the representation must also use the right concepts, so that its conceptual structure matches reality's structure. There is an objectively correct way to 'write the book of the world'.

Many contemporary philosophers of mathematics, starting with Benacerraf (1965), claim that mathematical structures are the subject matter of mathematics.

The debate between Plato and Aristotle on the relation between Forms and the sensible world is thus very much with us today, but it is now viewed through the prism of the notion of structure.

5.2 The Question of Realism

The realism debate in the philosophy of science started as a discussion about the extent to which we should believe existence claims on the strength of our best scientific theories, where it is recognised by all parties in the debate that our best theories evolve over time and will eventually be replaced by different, even better theories.

Realists say that we should believe in the existence of the unobservable objects, properties, and relations that are postulated by our best theories (McMullin 1984). Anti-realists say that we should only believe in the objects (properties, relations) postulated by our best scientific theories that can in principle be *observed* (van Fraassen 1980).

Over the past two decades, the realism debate has become more concerned with the *structure* that is posited by our best scientific theories. Worrall argued that it is rational to believe in the structures postulated by our best scientific theories, and that it is irrational to believe the claims made by our best scientific theories that are not structural (Worrall 1989). The motivation for this stance is the thought that the structural claims made by our best scientific theories tend to survive scientific revolutions, whereas non-structural claims of scientific theories are frequently overthrown in scientific revolutions. This position in the realism debate is called *epistemic structuralism*.

Ladyman and Ross (2009) go further than this and claim not only that only structure of reality can known but moreover that physical reality *consists* solely of structure. In particular, physical objects do not exist, according to their theory: they do not belong to the furniture of the universe. This position in the realism debate is called *ontic structuralism*.

This contemporary debate can be seen as a modern-day version of the debate between Plato and Aristotle. Poincaré and Worrall hold that the structure of the world is carried by objects and determined by their properties and relations. They just believe that anything that we can come to *know* about these objects, properties, and relations, is structural. This is an Aristotelian position. The position taken by Ladyman and Ross is a very strong form of platonism.

In the philosophy of mathematics after the Second World War we have witnessed a somewhat similar evolution.

Until well into the 1960s, the question of the existence of extra-mental mathematical objects was a focal point of the discussion in the philosophy of mathematics. In this discussion, we saw Gödel pitted against the more anti-platonist schools (versions of logicism, constructivism, formalism). But from the second half of the 1960s, the notion of structure came to play a more prominent role in the philosophy of mathematics in the Anglo-Saxon world.

Benacerraf (1965) argued that mathematics is about mathematical structures, and that there are no mathematical objects. He did not wish to deny (with Ladyman and Ross) that there are *physical* objects. Indeed, many of his followers have taken physical objects to be the carriers of the mathematical structures. And one need not even go as far as Benacerraf: one can take there to be mathematical objects *and* mathematical structures. But then mathematics is of course not *only* about mathematical structures.

This provided the context for a re-run of the debate between Plato and Aristotle in a new context: we can speak of platonism and Aristotelianism in the context of mathematical structuralism (Pettigrew 2008). Eliminative structuralists (or *in rebus* structuralists) claim that talk of mathematical structures is uniformly reducible to talk of objects and their relations: mathematical structures *supervene* on objects and their relations (Mayberry 1994; Burgess 2015). Thus we have arrived at an Aristotelian form of mathematical structuralism. Non-eliminative structuralists (or *ante rem* structuralists) do not deny that mathematical structures can be realised by systems of objects. But they hold that mathematical structures are *universals* that can exist independently of any instantiation in systems of objects. This is then a platonistic version of mathematical structuralism.

In this book I will be occupied only with *mathematical* structures. I will restrict myself to one particular context in which the concept of structure plays a role: *informal mathematical discourse*. And the concept of structure certainly does play a role in informal mathematical discourse: mathematicians speak of algebraic structures, geometric structures, discrete structures, and so on. Whereas outside mathematics we are mostly concerned with *embodied* existence of structure – we mostly speak of 'the structure *of*' – the language used by mathematicians does not immediately suggests that this is also the case in mathematics. Moreover, mathematicians appear to take mathematical objects to be situated in structures. The number 24 belongs to the structure of the natural numbers. The complete five-element simple graph (as a structure) contains five vertices.

In the spirit of naive metaphysics, I seek to develop a theory of the *nature* of mathematical structures. Also in the spirit of naive metaphysics, I am uninterested in the question of the possibility of *reducing* mathematical structure to mathematical or physical objects and their relations. So I will not be much concerned with the arguments pro and contra Aristotelian and platonist positions regarding mathematical structures. I will articulate a theory of the nature of structures as they present themselves to us and leave it at that.

I used to believe that mathematical theories are *about* mathematical structures, and was unsure about how mathematical objects fit in. Now I am not sure anymore that I quite understand the question: *what are mathematical theories about?* Any uniform metaphysical answer seems to me to go beyond mathematical practice, and by going beyond it, to distort it.

Some structuralists contend that mathematical theories are about mathematical structures and the positions in them, where the latter can be regarded as mathematical objects. Burgess (forthcoming) argues that structuralists go overboard if they say such things:

Mathematical practice refers to many sorts of objects that, even if they do not seem to be conceived as neo-logicist abstractions, equally do not seem to be referred to in connection with some background structure, let alone as 'structureless points' in such a background structure.

The point that Burgess is making here is that mathematicians routinely make essential use of *derived structure*.[2] But this derived structure does not strictly speaking belong to the 'basic structure' of interest.

To illustrate this, consider arithmetic. You will not get far into this discipline without being introduced to n-tuples of natural numbers. These n-tuples *can* be formally introduced by the method of taking equivalence classes ('abstraction'). But this is typically not how it is done in mathematical practice. Moreover, despite the fact that n-tuples of natural numbers can be *coded* as natural numbers (given some coding scheme or other), n-tuples *are* not strictly speaking positions in the natural number structure.

It will in any event not do to say that mathematical theories are *only* about mathematical structures and their objects, for it is also about processes (such as computations). Indeed, mathematical theories are also about concepts, questions, phenomena, etcetera. Therefore there seems to me no philosophically illuminating answer to the question what

---

[2] The distinction between basic and derived structure is discussed by Parsons (2004, p. 69). It can be traced back at least to the work of Manders (1989).

	Concrete	Abstract
Categorical	Arithmetic	Complex Analysis, Set Theory (?), …
Algebraic	Graph Theory, Computability Theory, …	Topology, Group Theory, Universal Algebra, …

Figure 5.1 A classification of mathematical theories.

mathematical theories are about. So I certainly do not want to make the broad claim that structure is the subject matter of mathematics.

I will be more than satisfied if I can shed some new light on the *nature* of mathematical structures and their objects. This will be the aim of the next chapters. Nonetheless, the present chapter is largely devoted to a discussion of eliminative and non-eliminative forms of structuralism. The reason for this is that both non-eliminative and eliminative versions of mathematical structuralism contain deep insights into the nature of mathematical structures: any satisfactory theory of mathematical structures will have to take these insights on board. What accounts of mathematical structures have not been able fully to do, in my view, is to connect these insights with a satisfactory theory of mathematical *object*.

There are two distinctions *between* mathematical theories that will turn out to be important for our discussion. Firstly, there are more *concrete* parts of mathematics and more *abstract* branches of mathematics. The more concrete parts of mathematics are those that are more closely connected to quasi-concrete objects. Thus arithmetic and (finite) graph theory are concrete branches of mathematics. Universal algebra and topology, on the other hand, are more abstract branches of mathematics. Secondly, there are *algebraic* and what one might call *categorical* branches of mathematics. Algebraic theories do not intend to describe one single mathematical structure but single out a family of structures. Categorical branches of mathematics, at least on the face of it, intend to describe one mathematical structure only. Group theory is an algebraic theory; real analysis is a categorical theory. For some fields – set theory is an example – it is controversial whether they are a categorical or an algebraic discipline.

These two distinctions give rise to a classification of mathematical disciplines, which is depicted in Figure 5.1.

5.3 Platonism and Reductionism

For the most part of the twentieth century, platonism was a minority position in the philosophy of mathematics. Its most famous proponent was Kurt

Gödel: he almost single-handedly revamped mathematical platonism and so made it a viable position again.

Gödel (1947) was a platonist both about mathematical objects and about mathematical concepts. He believed that mathematical objects and concepts are as objective as physical objects and properties. Mathematical objects and concepts are postulated in order to obtain a satisfactory theory of our mathematical experience. Indeed, in a way that is analogous to our perceptual relation to physical objects and properties, through mathematical intuition we stand in a quasi-perceptual relation with mathematical objects and concepts – and Gödel meant *all* mathematical objects and concepts, not just the quasi-concrete ones. Our perception of physical objects and concepts is fallible and can be corrected. Likewise, mathematical intuition is not fool-proof – as the history of Frege's *Basic Law V* shows – but it can be trained and improved. Unlike physical objects and properties, mathematical objects do not exist in space and time, and mathematical concepts are not instantiated in space or time.

One of the great foundational discoveries of the twentieth century is that all mathematical theories can be reduced to set theory, in the following sense. Every mathematical theory is about one or more entities each of which is isomorphic to a pure set equipped with some operations, and each of those is in turn isomorphic with a set. Take topology, for instance. Topology deals with topological spaces, each of which is isomorphic to some pure set. A topological space is isomorphic to the ordered pair $\langle A, B \rangle$, where A is a pure set, and B is a collection of subsets of A (the open sets of A), which forms again a pure set.

There is then a sense in which the subject matter of a mathematical theory *can* be seen as a collection of sets. If you are a platonist about sets, as Gödel was, then you are in a position to propose a form of ontological reductionism about all other mathematical entities. You can take all mathematical entities not just to be isomorphic to sets (which they are), but literally to *be* sets. This extra reductionist claim is not one that Gödel made. Nonetheless, in his philosophy of mathematics he concentrated on the nature and theory of sets, and he had little to say about any distinctive ontological nature that any other kind of mathematical objects (such as the natural numbers) might have.

Set theoretical reductionism was the target of the *identification problem* (also known as *Benacerraf's puzzle*) that was articulated in Benacerraf (1965). Benacerraf observed that for any mathematical entity (the natural numbers, any topological space, any Boolean algebra) there are *many* pure sets that it is isomorphic to. If there are two pure sets A and B that are both candidates for 'being' some antecedently given mathematical object E that is not obviously a pure set to begin with, then there seem to be no good reasons for thinking that E is identical with A rather than with B. Benacerraf claims in such situations that E is not identical to either A or to B.

Most philosophers have accepted the conclusion of Benacerraf's argument. A first possible reaction to this situation is to retreat to a form of platonism that recognises many kinds of mathematical entities (natural numbers, real numbers, complex spaces) to be not pure sets but collections of *sui generis* objects (the natural numbers, 'points' in a topological space). This is what most mathematical platonists appeared to be doing anyway. A more radical reaction, and one which Benacerraf seemed to favour, is to argue that any mathematical theory describes one or more *mathematical structures*, where a mathematical structure is something that is not ontologically reducible to sets. Categorical theories can then be said to describe one single structure, whereas algebraic theories describe a family of structures. But this raises the question: *What is a mathematical structure?*

5.4 Eliminative Structuralism

Trying to answer this question has led to a modern-day version of the debate between Plato and Aristotle. Let us first look at Aristotelian accounts of mathematical structure.

5.4.1 Eliminative structuralism and second-order logic. We can start by considering arrangements of objects that bear certain relations to each other. Let us call any such arrangement a *system* of objects. For simplicity and without essential loss of generality, take some such system $\langle S, R \rangle$, where S indicates the objects in question, and R indicates one relation that some of these objects bear to some of them. I am using set theoretic notation loosely here: I do not mean to appeal to the official notion of ordered pair when I use $\langle \ldots \rangle$, nor do I mean to imply that S is a pure or impure set. There are likely to be many other systems $\langle S', R' \rangle$ that stand in a structure-preserving one-to-one onto relation to $\langle S, R \rangle$. When such a relation between systems obtains, I will say that they are *system-isomorphic*, which is not to be confused with the set theoretic notion of isomorphism, since I have not assumed systems to be sets. The second system $\langle S', R' \rangle$ is then said to embody the *same mathematical structure* as the first system. Indeed, we are led to the following *abstraction principle for structures*:

Thesis 5.1. For any systems $\langle S, R \rangle$ and $\langle S', R' \rangle$:

$$\text{Structure of } \langle S, R \rangle = \text{Structure of } \langle S', R' \rangle \Leftrightarrow$$
$$\langle S, R \rangle \text{ and } \langle S', R' \rangle \text{ are system-isomorphic.}$$

This cannot be *completely* correct, as a Burali–Forti-like argument shows (Hazen 1985, pp. 253–254). In order to avoid this consequence, let us here restrict this principle to set-size systems.

The foregoing suggests that we may be able to say all we want to say about mathematical structures in terms of systems, i.e. without having to *reify* structures. In this sense, eliminative structuralism is an ontologically reductive programme. However, as Frege (1984, Section 66) reminded us a long time ago, an abstraction principle is not an explicit definition: it does not by itself allow us to eliminate the abstracted notion in every context. So we have check if we are able to state *enough* about structures without having to take them ontologically seriously.

Using a notion of *truth in a system*, we can express the truth conditions of arithmetical statement in terms of systems of objects. Consider an arbitrary arithmetical sentence φ. Restrict all quantifiers in φ to a second-order variable S. Replace all occurrences of the constant 0 in φ by a fresh first-order variable x_0. Replace all occurrences of $s, +, \times$ in φ by the second-order variables $R_s, R_+, R_\times$ (of suitable arity). Call the result of this translation procedure $\tau(\varphi)$. Likewise, for the principles of finitely axiomatised second-order Peano Arithmetic (PA^2): go through the same translation procedure, and call the result $\tau(\text{PA}^2)$. Then the truth conditions of φ are given in terms of systems in the following way:

$$\varphi^{Sys} = \forall S \forall R_s, R_+, R_\times \forall x_0 : \tau(\text{PA}^2) \to \tau(\varphi).$$

It is straightforward how this strategy can be applied to other mathematical theories.

The systems satisfying PA^2 are called *ω-sequences*. So according to this proposal, φ^{Sys} says that φ is true in all ω-sequences. Moreover, by a theorem due to Dedekind, all systems satisfying PA^2 are isomorphic (Truss 1997, p. 9). So this proposal does justice to the categoricity of arithmetic. This point generalises to other mathematical theories that we tend to think of as categorical, such as the theory of the complex field, for instance.

There is a difference of opinion among eliminative structuralists about how the content of the original statement φ relates to the content of φ^{Sys} (Burgess and Rosen 1997). Some say that φ^{Sys} makes explicit the content that φ had all along. This is called the *hermeneutic interpretation* of φ. Others hold that the content of the original statement φ differs from the content of φ^{Sys} but should be replaced by it. The latter view entails that we should change our ways. We should no longer use φ with the intention to be taken to have its original content (whatever that was). Rather, we should from now on intend to use it with its new content as given by φ^{Sys}. This is called the *revolutionary interpretation* of φ.[3]

For our purposes, there is no need to take sides in this debate. But there is a more pressing question. The above system-theoretic explication of the truth conditions of arithmetical statements appeal to second-order logic. But second-order logic is usually explicated in terms of the notion of *set* (Shapiro 1991). One of the recursive truth conditions for 'full' second-order logic says that a statement of the form $\forall X : \Phi(X)$ is taken to be true in a model $\mathfrak{M}$ if the interpretation of Φ holds for *every subset A* of the domain of $\mathfrak{M}$.

This means that the eliminative structuralist strategy is not immediately applicable to set theory itself: the strategy seems then not to eliminate the notion of set. In itself, this may not be problematic. However, there is a substantial assumption at work here. All noteworthy mathematical theories quantify either over infinitely many 'systems' or over systems consisting of an infinite number of objects (or both). So the interpretation of the system-theoretic translation of a mathematical statement will always involve reference to all subsets of an infinite collection. But the concept of *all subsets of an infinite collection* is one on which we do not have a firm handle. The history of the Continuum Hypothesis teaches us that even for the smallest infinite set, we do not know how many subsets it has, or even whether that question has a definite answer. So one may ask whether it is legitimate in an interpretation of mathematics to appeal to second-order logic in this way. The idea was always to appeal to logic in the interpretation of mathematical sentences. But it is not only a question whether second-order logic is logic: there is a question whether full second-order logical truth is a sufficiently determinate notion to be of interpretational use (Resnik 1988).

There is an alternative option: we do not *have* to interpret φ^{Sys} in the full second-order sense. We can just take φ^{Sys} to be a *two-sorted first-order statement*. This is exactly what Skolem (1922), for instance, would recommend. This amounts to quantifying over a wider class of models: all that is required is that they make the axioms of second-order logic true; no 'mystical' appeal is made in the metalanguage to quantifying over absolutely all subsets of a given infinite set. In effect, this results in admitting so-called

---

[3] There are more interpretations beside these two, but I leave them aside here.

non-standard models. For arithmetic, for instance, we would no longer be able to show that there is a unique structure that it describes. In general, from the point of view of *first-order* eliminative structuralism, the distinction between algebraic and categorical mathematical theories collapses: *all* mathematical theories are algebraic.

Many philosophers find the collapse of the distinction between algebraic and categorical theories an undesirable consequence of the first-order version of eliminative structuralism. There may be, however, a way of avoiding this consequence while still eschewing *all* set theoretic commitments. Natural language contains a device of *plural quantification*. In natural language we have a way of having a variable (indexical) refer to many objects at the same time, and quantifiers that range over all ways in which a variable can refer to many objects at the same time. As an example, take the sentence

Example 5.2. Some philosophers only cite each other.

This sentence cannot be formalised in first-order logic. It be formalised in the language of second-order logic as

$$\exists X [\forall y : Xy \rightarrow Py \wedge \forall x, y : (Xx \wedge Xy) \leftrightarrow (C(x, y) \wedge C(y, x))],$$

where $C(x, y)$ stands for the relation 'x cites y', and the predicate P expresses the property of being a philosopher. The quantifier-free matrix of this formula contains a second-order variable X. On the second-order logic interpretation, this variable ranges over *sets* of objects. On the *plural logic interpretation*, this variable ranges over pluralities. The plural interpretation seems preferable. Indeed, it is not clear at all why someone uttering the sentence in example 5.2 should thereby commit herself to the existence of sets!

It seems that one can explicate the semantics of the language of (monadic) second-order logic purely using plural quantifiers, without using the concept of set (Boolos 1985). Given the fact that in arithmetic we have a coding function available, in this particular context we can also explicate the semantics of the language of polyadic second-order arithmetic using only plural quantification.

This does not remove all of the concerns that we have discussed earlier. Resnik has defended the Quinean thesis (see Quine 1986, p. 66) that pluralities are only sets in disguise (Resnik 1988). The idea is that when we use plural anaphora such as 'them', we are putting several objects under one umbrella, and that is all that it takes for them to be a set. I will not express a judgement about the extent to which this objection is convincing. I do want to note, however, that the objection of entanglement with possibly inherently indeterminate problems such as the Continuum Hypothesis applies as much (or as little) to the plural quantification version of eliminative structuralism as for the full second-order version of eliminative structuralism.

5.4.2 Whence the systems? Let us return to the eliminative structuralist translation of arithmetical statements. One immediately sees that the translation procedure only gets the truth condition of arithmetical statements right *if there are infinite systems*. But if there are no infinite systems, then *all* arithmetical sentences come out *trivially* true under the translation function. So an infinite plurality of objects is required for this interpretation to work. And, as before, this holds not only for arithmetic but for every mathematical theory that postulates infinitely many objects.

Where can the infinite systems be found that ensure that the truth conditions of mathematical statements, as given by eliminative structuralism, come out right?

You might appeal to set theory. If you assume the iterative hierarchy of pure sets, then you have more than enough systems to get the eliminative structuralist truth conditions of statements of ordinary mathematical theories to come out right. Indeed, in any textbook of any mathematical field, the systems that the field deals with are typically in the first pages of the book *defined* to be sets with operations on them that satisfy certain conditions. This version of eliminative structuralism is called *set theoretic structuralism*. It was championed by the collective of mathematicians known as Bourbaki (Mayberry 1994).

Of course this means that you are according an exceptional status to set theory. In your account, set theory is interpreted not in a structural manner but in a straightforward platonist way. But many adherents of eliminative structuralism have physicalist leanings. For them, making use of the iterative hierarchy of sets in their structuralist account of mathematics is not an attractive way to go. So I will leave set theoretic eliminative structuralism aside from now on.

There is no immediate reason why the systems that the eliminative structuralist should appeal to must be set theoretic. I was careful in the beginning of this section not to define a system of objects as a *set*. Moreover, if we adopt a plural reading of the second-order quantifiers, then also our use of second-order logic does not contaminate our account with set theoretic commitments. So perhaps we can find the systems that we need in the physical world.

Hilbert (1930, pp. 380–381) already saw that this strategy may be problematic[4]:

Was den Begriff 'Unendlich' betrifft, so müssen wir uns klarmachen, daß 'Unendlich' keine anschauliche Bedeutung und ohne nähere Untersuchung überhaupt keinen Sinn hat. Denn es gibt überall nur endliche Dinge. Es gibt keine unendliche Geschwindigkeit und keine unentlich rasch sich fortpflantzende Kraft oder Wirkung. Zudem ist die Wirkung selbst diskreter Natur und existiert nur quantenhaft. Es gibt überhaupt nichts Kontinuierliches, was unendlich oft geteilt werden könnte. Sogar das Licht hat atomische Struktur, ebenso wie die Wirkungsgröße. Selbst der Weltraum ist, wie ich sicher glaube, nur von endlichen Ausdehnung. ... Das Unendliche ist nirgends realisiert; es ist weder in der Natur vorhanden noch als Grundlage in unserem Denken ohne besonderen Vorkehrungen zulässig.

In other words, our physical world may contain only finitely many objects. Even if it in fact contains infinitely many objects, this is probably only contingently so. But it seems that mathematical truth should not be held hostage to cosmological fortune. Moreover, even if the world contains infinitely many objects, they may not have the right physical structure to interpret mathematics. For instance, the existence of infinitely many physical objects related by a part-whole relation governed by the Quine–Goodman postulates of mereology does not suffice even to interpret elementary arithmetic (Niebergall 2000).

Field (1980) has argued that space (or spacetime) can be used to interpret mathematical theories. He argued that the regions of space carry physical structure that permits the interpretation of Real Analysis. An advantage of this proposal is that it is hard to imagine what it would be like for our physical world not to be spatial. So perhaps the necessity of mathematical truth is somehow safeguarded. But the situation is more complicated than it appears to be at first sight. It may indeed be difficult to imagine our world not being spatial. But the viability of Field's account depends on the correctness of his metaphysical theory of space. In particular, it presupposes a substantivist (as opposed to a relational) account of space. Moreover,

---

[4] 'As far as the concept 'infinity' is concerned, we have to be clear that 'infinite' does not have any intuitive sense and has no meaning at all in advance of closer investigation. For there are everywhere only finite entities. There is no infinite velocity and there are no forces or influences that can be transmitted with infinite speed. Moreover, force itself is discrete and it only exists in quanta. There is nothing that is continuous, that can be divided infinitely many times. Even light has atomic structure, as does the measure of force. Even the universe is, as I certainly believe, only finite in extent ... The infinite is nowhere revealed; it is neither given in nature, nor can it be admitted without special precautions as a basis of our thought' (my translation).

his proposal does not include a straightforward interpretation of parts of mathematics that appear to postulate the existence of many (more than 2^{2^ω}) mathematical objects.

An appeal to *modality* may be of use to the physicalist eliminative structuralist at this point. As observed earlier, it seems that there might have been fewer physical objects than there are. But, likewise, there might have been *more* physical objects than there are. In particular, our world might have contained infinitely many objects that are physically related in such ways as to allow the interpretation of significant parts of mathematics. This means that in particular the following is true (Hellman 1993):[5]

$$\lozenge \exists S \exists R_s, R_+, R_\times \exists x_0 : \tau(\mathrm{PA}^2).$$

What is more, this statement is *necessarily* true.

Moreover, let us strengthen the translation procedure for arithmetical statements in the following way:

$$\varphi^{Sys*} = \square \forall S \forall R_s, R_+, R_\times \forall x_0 : \tau(\mathrm{PA}^2) \to \tau(\varphi).$$

This is legitimate, for we want the truths of arithmetic to hold in all possible situations where there is a system that is rich enough to make the basic principles of arithmetic true.

The resulting proposal is called *modal structuralism*. It gets the truth conditions for arithmetic right and, moreover, allows us to make sense of the apparent *necessity* of arithmetical truth. Clearly a similar strategy can be applied to other mathematical theories. But for the higher reaches of mathematics it is not clear that it works well. As Parsons (1990) has pointed out, it is not clear, for instance, that the modal statement

$$\lozenge \exists S \exists R : \tau(\mathrm{ZFC}^2),$$

which should guarantee the non-triviality of (second-order) set theory (ZFC^2), is true. It is just difficult to see a vast possible universe as is required to make this statement true as in any way *concrete* or *physical*.

5.4.3 No mathematical objects. The abstraction principle for structures that was discussed in the beginning of this chapter (Thesis 5.1) allows us to talk about structures without taking structures to be ontologically

---

[5] In this formula, the sentential operator $\lozenge$ expresses possibility; its dual operator $\square$ expresses necessity.

fundamental. Categorical theories are said to be about individual structures, whereas algebraic theories are about *families* of individual structures that share certain general structural features. So individual mathematical structures are fundamental for mathematical structuralism (Isaacson 2011, pp. 1–2):

Modern mathematics is much more about general structures, but despite this shift, the reality of mathematics turns ultimately on the reality of particular structures ... The particular structures of mathematics constitute the determinate reality and objectivity of mathematics.

It is natural to go on to say that mathematical structures contain mathematical objects, and indeed that mathematical theories are also about mathematical objects (Parsons 2008, p. 1):

The language of mathematics speaks of objects. This is a rather trivial statement; it is not clear that we can conceive any developed language that does not. What is of interest is that, taken at face value, mathematical language speaks of objects distinctively mathematical in character: numbers, functions, sets, geometric figures, and the like. To begin with, they are distinctive in being abstract.

But in eliminative structuralism there is no room for mathematical objects (Shapiro 1997, p. 10). Tellingly, Hellman's monograph on modal structuralism bears the title *Mathematics without Numbers*.[6]

The idea of modal structuralism is that even if the world that we inhabit is radically finite – or even if Ladyman and Ross are right and it contains no physical objects at all – and therefore does not contain anything like the natural numbers, mathematics is not in need of revision. But more is true. On any eliminative structuralist position, even if our physical world contains enough physical objects (standing in suitable physical relations to each other), there are no numbers. In such a situation, any physical object can play the role of the number 51, but so can any other physical object, so *the* number 51 does not exist. Isaacson (2011, p. 2) sums up the position of the eliminative structuralist neatly:

The question whether mathematical objects exist is misguided. We have truth and realism without reference in mathematics.

---

[6] This is a pun on the title *Science without Numbers* of Field (1980).

Indeed, on the eliminative structuralist's view, even to speak of *the* natural number structure is misleading. There *really* are only systems of objects, and none of them is the natural number structure.

If there are no mathematical objects, then Benacerraf's identification problem evaporates. If there are no individual natural numbers, and there are no finite ordinals (as defined set theoretically in the standard way), then there simply is no question whether $2 = \{\emptyset, \{\emptyset\}\}$.

Nonetheless, we do ordinarily speak of 'the number 1113'. Eliminative structuralists have developed a semantic theory to account for this. Pettigrew has argued that (most) mathematical definite descriptions function as *dedicated variables*. The idea is that in non-mathematical natural language as in mathematics we use both ordinary variables and dedicated variables. Ordinary variables in natural language are expressions that take on different values in different contexts. The word 'it', for instance, can refer to a particular fire truck in one context, and Chaminade's *Opus 89* in another context. Dedicated variables, in contrast, are governed by the convention that once they have been assigned a value, it 'sticks to them': they are then not permitted to be used later on for referring to something else.

An example from natural language of a dedicated variable is the expression 'Tommy' in the following slogan from the Second World War (Pettigrew 2008, Section 2.2.3):[7]

Example 5.3. Tommy needs his letters from home.

Concerning arithmetic, the thought is that the expressions '12', 'ℕ', '+', ... have all been introduced simultaneously by coordinated meaning conventions. These expressions have been introduced as dedicated variables, so that their value is required to be the same wherever they occur and coordinated with the other arithmetical dedicated variables. Moreover, in an eliminative structuralist spirit, these variables are assumed to be implicitly universally quantified over 'from the outside'.

Of course this is not the only possible semantic theory of 'Tommy needs his letters from home'. Indeed, in the light of Chapter 3, one might ask we cannot take the term Tommy to refer to an *arbitrary* British soldier and thus take it really to be a name. This already suggests that perhaps the number 1113 can be taken to be an arbitrary object too.

---

[7] The number of non-scientific natural language examples that one could list here appears to be somewhat limited.

5.5 Non-eliminative Structuralism

I now turn to non-eliminative forms of structuralism. The focus of the discussion will mostly be on Shapiro's version of non-eliminative structuralism.

5.5.1 *Ante rem* structures. We have seen that for the mathematical structuralist, individual mathematical structures are fundamental. The non-eliminative structuralist takes apparent wisdom about mathematical structures, as contained in what mathematicians informally say about structures, seriously. She believes that the philosopher of mathematics should attempt to save these appearances. We have seen that in her account of names and descriptions, the eliminative structuralist diverges from appearance: what appear to be names ('12') are not really names, what appear to be definite descriptions ('the natural number structure') are not really definite descriptions. In Shapiro's view, we should try to do better.

Shapiro accepts Benacerraf's contention that designating some particular ω-sequence as *the* natural number structure is hopelessly misguided. And he also accepts the eliminative structuralist's claim that particular systems exemplify (embody, instantiate) the natural number structure.

What Shapiro proposes is that the natural number structure can be obtained by *abstraction* from the idiosyncracies of the exemplifications of the natural number structure. *The* natural number structure is what all systems that exemplify the right structure have in common with each other, where exemplifying the right structure is – in agreement with eliminative structuralism – making the principles of PA^2 true. The natural number structure is not a set. Instead, it is a *sui generis* entity: it is a *structural universal*.

The natural number structure thus consists of *positions* (places, offices, roles) that bear appropriate structural *relations* to each other. These positions can be occupied by particular objects that have intrinsic properties and internal structure. But the positions themselves have no internal properties or structure: for any position, objects with different internal structures can occupy that position (Resnik 1981, p. 530):

In mathematics, I claim, we do not have objects with an 'internal' composition arranged in structures, we have only structures. The objects of mathematics . . . are structureless points or positions in structures. As positions in structures, they have no identity or features outside a structure.

Shapiro claims that the natural number structure is ontologically independent of all possible concrete systems that can instantiate it. By extension, every position in that structure is ontologically independent of all particular objects that can occupy that position.[8] The natural number structure is therefore an *abstract* entity: we have arrived at a *platonist* version of structuralism.

In Shapiro's view, the positions in the natural number structure can themselves be regarded as *objects*. Thus the natural number structure can be regarded as a plurality of objects that are related to each other in certain ways. In other words, the natural number structure is itself a *system*. Not only that, but it is a system with the 'right' structure: *the natural number structure instantiates itself* (Shapiro 1997, p. 89):

Structures exist whether they are exemplified in a non-structural realm or not. On this option, statements in the places-are-objects mode are taken literally, at face value. In mathematics, anyway, the places of mathematical structures are as bona fide as any objects are. So, in a sense, each structure exemplifies itself. Its places, construed as objects, exemplify the structure.

One might ask under which circumstances a position in one structure is identical with a position in another structure.[9] Resnik (1981) argued that such questions are *meaningless*: only intra-structural questions of identity and difference make sense. Shapiro (1997) initially agreed with him. But he later argued that all identity statements between positions of different structures are *false* (Shapiro 2006).

At this point, Shapiro considers the job done. He believes that he has found a way to do justice to Benacerraf's identification worries but at the same time respect the semantic appearance of mathematical terms (Shapiro 1997, p. 11):

The ante rem *structuralist interprets statements of arithmetic, analysis, set theory, and the like, at face value. What appear to be singular terms are in fact singular terms that denote bona fide objects. Moreover,* ante rem *structuralism accommodates the freestanding nature of mathematical structures. Anything at all can occupy the places of the natural-number structure, including natural numbers themselves. Thus, I hold that* ante rem *structuralism is the most perspicuous account of contemporary mathematics.*

---

[8] Note, incidentally, that this claim does not logically follow from taking structures to be *sui generis* entities that exist over and above the concrete systems that exemplify them. Someone might claim that nonetheless structures are ontologically dependent on the systems that exemplify them.

[9] This question is discussed in MacBride (2005).

The subject matter of an algebraic theory is then not a single free-standing structure, but a *family* of such structures (Shapiro 1997, p. 73 n2).

Shapiro (1997) applies his non-eliminative structuralism to set theory in the following way. The subject matter of set theory consists of all the sets there are. These sets are *sui generis* structures, consisting of places ('elements') with a relation on them (the 'elementhood' relation) (pp. 103–104). These places do not have any internal structure, so one does not go far astray if one pictures them as directed graphs. The metaphor of a set as 'a plurality of objects taken as one' may play a role in the motivation of axioms for set theory, for it gives us some confidence that some axioms for set theory are jointly consistent when thus interpreted. But in the official account of what set theory is about, the metaphor cannot play a role. The sets, taken as *ante rem* structures, do *not* jointly form one great 'mother structure': the structural counterpart of the platonist's universe of sets (Shapiro 1997, pp. 95–96). Rather, they form a mere potential infinity of finite and infinite structures (Shapiro 1997, p. 202). This means that Shapiro ultimately regards set theory as an algebraic discipline.

Shapiro's non-eliminative structuralism is not able to take *all* informal mathematical talk about structures completely literally. Some of it is, when taken literally, simply incoherent, and it is not *meant* to be taken literally. Here is an example. A graph theorist will indeed talk about *the* complete two-element graph, and a non-eliminative structuralist can take this literally. But the graph theorist will also say that the complete three-element graph contains *three* complete two-element graphs as subgraphs. The non-eliminative structuralist will say, quite sensibly, that the latter assertion can be *paraphrased* as follows: every system that instantiates the complete three-element graph contains three subsystems that instantiate the complete two-element graph.

Given that in Shapiro's view the natural number structure instantiates itself, the claim that the natural number structure is ontologically independent of all its instantiations is not quite correct: clearly the natural number structure is not ontologically independent of *itself*. But the natural number structure will still be ontologically independent of all possible physical ω-sequences.

5.5.2 Objections. Physicalists will of course object to the platonist flavour of Shapiro's non-eliminative structuralism. Any mathematical structure differs from every concrete system that instantiates it. This means that mathematical structures are abstract. Has not Science taught us that abstract entities do not exist?

I do not believe that science has taught us any such thing. But be that as it may, the question of the existence of abstract entities has been with us at least since Plato, and it is not going to resolved soon. So I will leave this issue aside: it is too big for me to deal with. Instead, I concentrate on arguments to the effect that Shapiro's non-eliminative structuralism is also subject to problems that are quite independent of any possible distaste of abstracta.

The objects that populate *sui generis* structures are according to non-eliminative structuralism in some sense *incomplete*. A rough statement of what is intended (when applied to arithmetic) is to say that numbers only have *structural properties* (Shapiro 1997, pp. 72–73). This allows Shapiro to give at least a partial answer to Benacerraf's puzzle. If you were to ask, for example, whether the number 13 of the natural number structure is the same object as the number 13 of the structure of the rational numbers, the answer would be that there is no fact of the matter or that the answer is trivially no. Only questions that pertain to the role of the natural number 13 in the structure of the natural numbers have interesting answers.

This sounds attractive. However, it has turned out to be very difficult to make the intended meaning of such statements sufficiently precise in a way that does not lead to counterexamples.[10] For instance, the number 7 might have the property of being my least favourite number, even though this is not a 'structural' property.

Burgess (1999) has pointed out that beside this problem of how positions in a structure relate to objects outside the structure, the non-eliminative structuralist conception of the incompleteness of positions gives rise to questions about how positions inside structures are differentiated from each other (pp. 287–288):

In the case of the natural numbers for instance, ...though they are supposed to have no non-structural properties, at least each has a structural property, expressible in the relevant first-order language, that distinguishes it from the others: only one of them comes first in the ordering on natural numbers, only one comes next-to-first, only one comes next-to-next-to-first, and so on – and that's all the natural numbers there are ... The situation changes, however, when we come to the complex numbers. There we have two roots to the equation $z^2 + 1 = 0$, which are additive inverses to each other, so that if we call them i and j we have $j = -i$ and $i = -j$. But the two are not distinguished from each other by any algebraic properties, since there is a symmetry or automorphism

---

[10] Shapiro himself has also come to recognise that it is difficult to give a satisfactory philosophical account of the incompleteness of mathematical objects (Shapiro 2006, Section 1).

of the field of complex numbers, ... which switches i and j. On Shapiro's
view the two are distinct, though there seems to be nothing *to distinguish*
them ... Shapiro offers no extended discussion of the mystery of symmetry, and
I consider this the most serious omission in the book.

Keränen (2001) later went further and argued that this mystery cannot be
solved, and took this as evidence that non-eliminative structuralism should
be rejected.

To conclude, Hellman (2006, p. 546) points out that Shapiro's view is
vulnerable to a *permutation objection*. If, in the *ante rem* structure $\mathbb{N}$ of
the natural numbers, we permute its places (in a non-trivial way), then we
obtain a system **N** that is isomorphic to $\mathbb{N}$. We can then ask, in the spirit
of Benacerraf (1965): what could possibly make $\mathbb{N}$, rather than **N** be the
unique *sui generis* structure that arithmetic is about? In other words, there
is a worry that Shapiro's position is vulnerable to a version of Benacerraf's
puzzle.

I will not assess these objections at this point. Instead, I will return to
them in Chapter 8, where I try to develop a philosophical view of mathematical structures that gives a better account of the incompleteness of
mathematical structures and mathematical objects.

5.6 Structure and Computation

Eliminative structuralists propose that a discipline such as arithmetic is
about many systems at the same time, so to speak. The question then arises
which conditions a system has to satisfy in order to belong to the class
of systems that arithmetic is about. Let us call systems that meet these
conditions *admissible systems*.

Benacerraf addressed the question what the conditions on admissible
systems are. He argued that for a system $\langle S, <, 0, +, \times \rangle$ to belong to this
class, the following must hold (Benacerraf 1965, p. 53):

1 $\langle S, <, 0, +, \times \rangle$ must satisfy the principles of first-order arithmetic.
2 $\langle S, < \rangle$ should be an ω-sequence.
3 $<$ should be a recursive relation on S.

The requirement of $\langle S, < \rangle$ being an ω-sequence, in combination with
the requirement that $\langle S, <, 0, +, \times \rangle$ satisfies the axioms of first-order arithmetic semantically do the same job as being a full second-order model of
PA^2. Any two systems that make PA^2 true are system-isomorphic. So from

a model-theoretic point of view these two systems are completely indistinguishable. From this perspective, there are no grounds for preferring one over the other.

From the model-theoretic perspective the recursiveness requirement is not even well defined, because recursiveness, as officially and formally defined in the textbooks on computability theory, is defined on the natural numbers, whereas the underlying pluralities of many ω-sequences do not contain numbers. But Benacerraf does have a point, which is expressed well by Shapiro (1982, p. 14):

Mechanical devices engaged in computation and humans following algorithms do not encounter numbers themselves, but rather physical objects such as ink marks on paper. Since strings are the relevant abstract forms of these physical objects, algorithms should be understood as procedures for the manipulation of strings, not numbers. Furthermore, mathematical automata, such as Turing machines, which are the abstract forms of computation devices, have only appropriately constituted strings for inputs and outputs. It follows that, strictly speaking, computability applies only to string-theoretic functions and not to number-theoretic functions.

This holds a fortiori for the eliminative structuralist who believes that the numbers themselves do not exist. So the notion of recursiveness that Benacerraf is using is an *informal* notion (Benacerraf 1996, p. 1986 n5), and our informal notion of computability is a *practical* one.

The sense in which Benacerraf (1965, pp. 51–52) thought that the recursiveness requirement on $<$ is still necessary, is the following. One thing we must be able to do with the natural numbers, and we *learn* to do with them at a tender and impressionable age, is to *count them out* from small to large. Now take a highly noncomputable subset N_0 of the natural numbers. N_0, ordered from small (in $\mathbb{N}$) to large (in $\mathbb{N}$) by the relation $<_{N_0}$ forms an ω-sequence. So if $0_{N_0}, +_{N_0}, \times_{N_0}$ are defined on N_0 are defined appropriately, then $\langle N_0, <_{N_0}, 0_{N_0}, +_{N_0}, \times_{N_0} \rangle$ is an isomorphic copy of the standard model $\mathbb{N}$. But we should be able to count out the elements of N_0, and we can't. So $\langle N_0, <_{N_0}, 0_{N_0}, +_{N_0}, \times_{N_0} \rangle$ is not a system that arithmetic can be taken to be about.

At this point you might think that this example is not relevant to eliminative structuralism in any case. The reason is that the example appeals to the natural numbers, and for the eliminative structuralist there are no natural numbers. But this is not of the essence. We can surely imagine possible *physical* systems that fully satisfy the axioms of PA^2 but have a 'smaller than' relation that is by no means informally computable.

The philosophical literature about arithmetical structuralism has mostly ignored Benacerraf's recursiveness requirement. And Benacerraf came to think that they were right to do so: he realised that his argument for the recursiveness condition was not a good one. Our good old arabic numerals can be taken to count out the elements of N_0. Indeed, who knows, perhaps they do! Benacerraf (1996, p. 188) put the point as follows:

Someone who was speculatively inclined, reflecting on our present situation, might even imagine that our familiar canonical notations were similarly the products of divine gerrymandering, designed for the purpose of affording us ready and convenient 'access' to their designate. Happily, on this scenario, our unary, binary, octo, decimal, Roman, etc. systems (as well as all others recursively intertranslatable with them), ...reveal the order of the numbers, all while forming a notational veil of innocence that masks their true natures.

In other words, on the eliminative structuralist picture, when we say 'zero, one, two, three', we automatically count out the numbers of *all* the ω-sequences, no matter how uncomputable their ordering relation.

Benacerraf was concerned about *counting* and deciding the ordering relation on admissible systems. But more in general, a philosophical account of arithmetic should contain an account of *computation*, i.e. of what it means to compute functions on the natural numbers. So we need an account not only of the computability of the ordering relation, but also of elementary functions on the natural numbers such as $+$ and $\times$.

According to the textbooks of recursion theory, computations take natural numbers (or n-tuples of them) as inputs and yield natural numbers as outputs. But the discussion so far has revealed that according to arithmetical structuralism, this cannot literally be correct. According to eliminative structuralism, natural numbers are positions that have no internal structure. But in computation essential use is made of the internal structure of the entities that are computed with. For eliminative structuralism there are no natural numbers, so it is even more immediate that according to this theory computation on the natural numbers makes no literal sense. There are admissible systems **A** such that we can practically compute on and with the elements of the domain of **A**. But these form only a sub-collection of the admissible systems. So for eliminative structuralism it is not even a requirement that we should be able to compute on the entities that can play the role of the natural numbers. Nonetheless, they will insist that by means of our familiar informal computation procedures on the arabic numerals, strings of zeros and ones, and so on, we are able to compute the appropriate class of functions of every system.

Now imagine that you have *formalist* sympathies. Suppose you believe, with Hilbert perhaps,[11] that notation systems are privileged in the sense that arithmetic is first and foremost *about* such systems.[12]

Let us take a *computable notation system* to be a system $\langle N, s, +, \times \rangle$ (in the sense used in Section 5.4.1) satisfying the principles of first order Peano Arithmetic, where its domain N is an arithmetical notation system N that is computably generated, and its operations $s, +, \times$ on N are informally computable. Examples of such systems are the system of arabic numerals, the binary system, and the system of roman numerals, all with the elementary functions $s, +, \times$ defined on the notations in the natural way. You might want to restrict the *admissible* systems to systems that we can actually compute on, i.e. to computable notation systems. The variant of eliminative structuralism that does this is called *computational structuralism* (Halbach and Horsten 2006; Horsten 2012). Then Tennenbaum's theorem ensures that we still have categoricity: any two such computational notation systems are isomorphic (Kaye 1991, Chapter 11). In fact, the admissible models according to computational structuralism are exactly the models that belong to the recursive isomorphism type generated by the admissible system of the arabic numerals. In sum, the upshot is that computational structuralism has pinned down the right structure, and has forged a tighter connection between admissible systems and computation than is presumed by eliminative structuralism for arithmetic in general.

All of this is about arithmetic and closely related structures (such as the structure of the rationals) because these can be taken to be exclusively about quasi-concrete entities. For a highly abstract discipline such as topology it is not clear whether the considerations of this section have any relevance at all. But for real analysis, they do have some relevance. One might ask whether there is a Tennenbaum phenomenon for the reals: are there reasonable computability restrictions on models of real analysis that pin the class down to the isomorphism type generated by the standard model of analysis? As far as I know, this is at present an open question.

The conclusion of the present section is *not* that computational structuralism for arithmetic is superior to standard eliminative structuralism for arithmetic. Indeed, I won't pursue the question which form of arithmetical

---

[11] '[A]m Anfang ... ist das Zeichen' (Hilbert 1935, p. 163). ('In the beginning ... there is the sign' [my translation]).

[12] Shapiro (1982) contains an investigation of the relation between notation systems and computation.

structuralism is the most attractive one. I do claim that the class of computable models of arithmetic is a natural and interesting one. It will be of relevance in the next chapter.

5.7 The Peculiar Case of Set Theory

At the most concrete end of the spectrum of mathematical theories, we find arithmetic. We have seen that its connection with computation poses special questions for mathematical structuralism. At the other far end of the spectrum, we find set theory. But set theory is more than one of our most abstract mathematical disciplines: it resists being seen through a structuralist lens.

In the discussion of set theory vis-á-vis mathematical structuralism, let us put impure sets aside, and concentrate on the theory of pure sets.

According to the 'naive' platonist view, set theory is about the iterative hierarchy of pure sets. So, on this view, set theory is a categorical discipline.

Moreover, formal categoricity arguments, analogous to Dedekind's categoricity arguments for arithmetic and analysis, have been produced for set theory. Zermelo argued in full second-order logic for the *quasi-categoricity of set theory*, by showing that for any two systems that satisfy ZFC^2, either they are isomorphic, or one of them is isomorphic to an initial rank of the other (Zermelo 1930). Attempts to strengthen this to categoricity argument have later been made. Donald Martin (2001), for instance, has argued that up to isomorphism there is only one full system of set theory. This is connected to a research programme in mathematical logic going back to Gödel (1947), which is known as the *large cardinal programme*. This programme aims at reducing the incompleteness of our understanding of the one and only set theoretic universe by uncovering ever stronger principles of infinity that hold of *the* set theoretic universe.

Many philosophers of mathematics and mathematical logicians do not find categoricity arguments for set theory persuasive because of their inevitable use of some degree of higher-order logic. Indeed, many take – like Shapiro – set theory to be an algebraic discipline. They often argue that the history of the technique of forcing has taught us that there is a plethora of non-isomorphic systems that make the principles of ZFC true and that are just as good and natural as any models of ZFC (Hamkins 2012). This view has led to a research programme in mathematical logic that aims at giving a mathematical description of the *set theoretic multiverse*.

How does this bear on our discussion of mathematical structuralism?

If you accept the 'naive' platonistic view about set theory, or the arguments that intend to show that set theory is about one unique system, then there clearly is no room for structuralism about set theory. The set theoretic universe V is a *rigid* structure: it admits no non-trivial automorphisms. If you accept Zermelo's quasi-categoricity argument (and reject all categoricity arguments that intend to establish a stronger conclusion), then set theory must be situated in the intermediate area between categorical and algebraic theories. On this picture, you can be a structuralist about sets. The transfinite ranks that satisfy ZFC^2 can then be seen as the systems that set theory is about, and no overarching universe of sets of which all these full second-order models are parts needs to be countenanced (Isaacson 2011).

Perhaps set theory does occupy a unique position in the discussion of mathematical structuralism. Burgess (2015, p. 144) has argued that the *motivation* for taking a structuralist position seems absent in the case of set theory, for at first sight it does not appear to be the case that we can easily find many systems that serve equally well as our set theoretic universe. Of course this is exactly what the defenders of the set theoretic multiverse hypothesis deny: they argue that the set theoretic multiverse is a perfect setting for an eliminative structuralist view about sets. Moreover, we have seen above how it can be argued, in the spirit of Zermelo, that even if you accept full second-order logic, there is room for eliminative structuralism about sets (Isaacson 2011). We saw in Section 5.5.1 that Shapiro holds a position along these lines.

There is another consideration in support of the thesis that set theory has a special status in the debate about mathematical structuralism.

Some believe that the concept of set is unclear (Oliver and Smiley 2006). It is difficult to go beyond metaphors (of 'collecting objects', of 'throwing a lasso around some objects') in explicating the concept of set. Moreover, it is sometimes considered especially unclear what the empty set is ('a collection with nothing in it'), and some find it very difficult to see how the singleton containing the element *a* (and the element *a* only) differs from *a* itself.

Nonetheless, I submit that we have a *much* clearer idea what sets are than, for instance, what natural numbers are. Our concept of set is fundamental, so no one should expect that metaphors will ever be able to convey the concept of set completely. But the concept of set is more than a 'purely structural' concept. Positions ontologically depend on the structure to which they belong, and vice versa. But our concept of set entails that sets ontologically depend on their elements, whereas elements do not ontologically depend on the sets to which they belong (Parsons 2008, p. 118). Moreover, the perceived problems concerning the empty set and the singletons are problems at the fringes, so to speak: they do not show that the concept of set in general is unclear.

We have seen in Section 5.4.3 how Parsons emphasises that it is difficult to regard infinite systems in general, and transfinite sets in particular, as even possibly *concrete*. This might be seen as an argument for the thesis that transfinite sets are really *sui generis* structures, perhaps in the sense of Shapiro (Parsons 2008, p. 124). One can go even further, and take the sets to form one gigantic free-standing structure. The point that I want to make is that there is not much motivation for going down this road. There is no obvious counterpart for sets of Benacerraf's identification problem for natural numbers. Sets do not seem to have the sort of 'incompleteness' that the natural numbers, for instance, do seem to have (see Section 5.5.2). Moreover, as argued above, it is much clearer what it means to be a set than what it means to be a number.

6 | Mathematical Structures

Many philosophers of mathematics today believe that mathematical structuralism is broadly correct. Yet there is much disagreement between mathematical structuralists. Much of this internal disagreement revolves around problems of *realism* about mathematical structures. The question that has dominated the literature is whether mathematical structures are fundamental, i.e. whether they belong to the basic furniture of the world. Thus much of the discussion on mathematical structuralism is situated in foundational metaphysics.

This has stood in the way of deepening our understanding of the *metaphysical nature* of mathematical structures. In this chapter and the next, we will put questions of realism of mathematical structures aside, only to return to them in Chapter 8. In the present chapter I develop a view of the nature of mathematical structures that differs from those in the literature, although it contains elements of existing forms of eliminative and non-eliminative structuralism. In Chapter 9 one structure is investigated in more technical detail: the structure of the natural numbers.

6.1 From Arbitrary Objects over Arbitrary Systems to Generic Systems

I propose a theory of mathematical structure that views the incompleteness of mathematical objects as a form of *arbitrariness*. Thus I will appeal to Kit Fine's theory of arbitrary objects as a framework for my account of mathematical structures.

Arbitrary numbers serve for Fine as a paradigmatic example of arbitrary objects. But the way in which what it means to be an arbitrary natural number was explained in Section 4.1 presupposes *specific* natural numbers as ontologically prior. Thus it is not immediately clear how the theory of arbitrary numbers fits with arithmetical structuralism.

We have seen in the previous chapter how the notion of *system* is a fundamental concept in mathematical structuralism, both in its eliminative and in its non-eliminative incarnations. Moreover, in non-eliminative structuralism a structure is an entity that is or *can be* exemplified by many systems.

I propose that we start likewise with the notion of a system as an arrangement of objects. Just as an arbitrary object is an entity that can be in the state of being a specific object, so an *arbitrary system* is an arbitrary entity that can be in the state of being a specific system, i.e. an arrangement of objects.

The state space of an arbitrary system may consist of systems that are all isomorphic with each other. In that case we say that the arbitrary system is a *generic system*. Generic systems will turn out to be 'rich'. They generally contain many arbitrary objects. Even mathematical objects are embedded, often in a canonical manner, as arbitrary objects in generic systems.

I claim that mathematical structures are generic systems, and vice versa. Within mathematical objects thus conceived, we can find *mathematical objects* that exhibit the right kind of incompleteness.

In the next sections, I will apply these ideas to a few examples. I will start with a discussions of fair dice and systems of them, before moving on to describe how the natural number structure can be seen as an example of a generic system.

Set theoretic models of aspects of the philosophical account that I propose will play an important role in what follows. They are helpful because they are precise enough to capture the mathematical content of the philosophical approach. Thus they allow us to recognise *consequences* and *properties* of my philosophical account that are not immediately obvious. Nonetheless, it has to be born in mind, and it cannot be stressed often enough, that giving a model is not the same as proposing an ontological reduction. Indeed, this chapter and the next are an exercise in *naive metaphysics*. The aim is to obtain a deeper insight in mathematical structures by any helpful means (even if that involves making use of set theory to provide models), without endorsing any ontological reduction of mathematical structures.

6.2 Fair Dice

Consider the *fair die*. It is an abstract object. The fair die is an arbitrary object. It can be in any one of six states with an equal probability ($\frac{1}{6}$). Any one of these states is itself also an abstract entity.

Consider now an arrangement of two fair dice. This is also an arbitrary entity, but it is an *arbitrary system* rather than an arbitrary object. Its state space consists of 36 states, and it can be in each of these states with equal probability $(\frac{1}{36})$. Each state of this system is a system consisting of two abstract objects. But the arbitrary system is not a *generic system*, for not all the states of the arbitrary system are isomorphic with each other.

Frequencies of outcomes in long sequences of throws with ordinary physical dice can be modelled quite well by reasoning about fair dice conceived of in this way. If the physical dice are carefully constructed to be as uniform in density and as symmetric as possible, then frequencies of outcomes will in long series of throws with these dice with overwhelming likelihood be close to $\frac{1}{36}$.

Thus arbitrary object theory gives a straightforward interpretation of some of the most elementary elements of probability theory. We see that the system of two fair dice is not a mathematical structure. But the system of two fair dice belongs to the subject matter of probability theory, which is a part of mathematics. This shows that mathematics is not *only* concerned with mathematical structures.

Let us compare this with a well-known passage in Kripke's (1980, pp. 16–17) *Naming and Necessity*, which is directed against Lewis' counterpart theory:

Two ordinary dice ... are thrown, displaying two numbers face up. For each die, there are six possible results. Hence there are thirty-six possible states of the pair of dice, as far as the numbers face up are concerned, though only one of these states corresponds to the way the dice will actually come out. We all learned in school how to compute the probabilities of various events (assuming equiprobability of states) ... Now in these school exercises in probability, we were in fact introduced at a tender age to a set of (miniature) 'possible worlds'. The thirty-six possible states of the dice are literally thirty-six 'possible worlds', as long as we are (fictively) ignoring everything about the world except the two dice and what they show ... Now in this elementary case, certain confusions can be avoided ... The thirty-six possibilities, the one that is actual included, are (abstract) states of the dice, not complex physical entities. Nor should any elementary school pupil receive high marks for the question 'How do we know, in the state where die A is six and B is five, whether it is die A or B which is six? Don't we need a 'criterion of transstate identity' to identify the die with a six – not the die with a five – with our die A? The answer is, of course, that the state (die A, 6; die B, 5) is given as such (and distinguished from the state (die B, 6; die A, 5)). The demand for some further 'criterion of transstate identity' is so confused that no competent schoolchild would be so perversely philosoph-

ical as to make it. The 'possibilities' simply are not given qualitatively ... If they had been, there would have been just twenty-one distinct possibilities, not thirty-six.

What Kripke says in this passage is perfectly compatible with taking the pair of fair dice to be an arbitrary system. Kripke starts by introducing a pair of ordinary physical dice, rather than about a pair of fair dice. The probability of *this* pair of dice coming up (die *A*, 6; die *B*, 5) will then not be exactly $\frac{1}{36}$, because, given that we live in an imperfect world, both dice will be slightly biased. So Kripke's assumption of equiprobability may be taken to imply that the pair of dice that we are concerned with is a pair of fair dice, which do not exist as physical entities in our world. Kripke then says that the 'possible outcomes' of throwing the dice are given by *mini-worlds*. These mini-worlds are not metaphysically possible worlds; but they are analogous to possible worlds (Martens 2006, p. 581). The analogy teaches us that just as *the very same* dice *A* and *B* can exist in different (and qualitatively indistinguishable) mini-worlds, so the very same person belongs to different metaphysically possible worlds – as long as 'metaphysically possible world' is given an appropriate deflationist reading.

6.3 The Generic ω-Sequences

Consider all the objects there are. To keep matters simple, suppose they form a countable infinity $a_0, a_1, a_2, \dots$, which I will refer to as *A*.

We have seen in Section 5.4.1 that a way of ordering *A*, or some of the *A*'s, into a well-ordering of type ω is a *system*: it is a specific ω-sequence. There are then 2^ω such.

These ways of ordering *A* into an ω-sequence are the states in which a *generic* ω-sequence can be. Borrowing and abusing terminology from classical mechanics, we may say that all ways of ordering *A* into an ω-sequence together constitute a *full state space* $\mathcal{S}$.

So in our supposed situation there are not just specific objects, and specific ω-sequences, but there are also generic systems: generic ω-sequences. Indeed, there are many generic ω-sequences. Generic ω-sequences are entities that *can be in the state* of being a specific ω-sequence. As we have seen to be the case with arbitrary objects in general (Section 3.3), it makes no sense to ask in which state a generic ω-sequence *actually* is.

In Section 4.1 the concept of being a *fully arbitrary* natural number was introduced. Similarly, there are *fully generic* ω-sequences. These are ω-sequences that can be in any state of $\mathcal{S}$. At the other end of the spectrum, there are the *minimally* generic ω-sequences. These are the ω-sequences such that there is exactly one state in which they can only ever be. So there is a sense in which the specific ω-sequences are canonically represented among the generic ω-sequences.

We also saw in Section 4.1 that arbitrary natural numbers are correlated with each other. Similarly, generic ω-sequences are correlated with each other. For instance, given some fully generic ω-sequence f, there is a fully generic ω-sequence g such that whenever f is in some state s, the generic ω-sequence g is in some other state.

Borrowing some more terminology from classical mechanics, we may take a point in the *configuration space* $\mathcal{C}_S$ generated by the generic ω-sequences to be a simultaneous assignment of states to all the generic ω-sequences. So the configuration space $\mathcal{C}_S$ specifies all possible situations that our supposed world can be in, as far as the generic ω-sequences are concerned: they can be thought of as 'possible worlds'.

It is not a priori determined how many elements this configuration space

$$\mathcal{C}_S = \{\mathcal{C}_1, \mathcal{C}_2, \ldots, \mathcal{C}_\beta, \ldots\}$$

has. But suppose we appeal to a principle of ontological sparsity according to which there are as many states in configuration space as are needed for having arbitrary ω-sequences that take all specific ω-sequences as values. Then, since there are 2^ω specific ω-sequences, we will also have that $|\mathcal{C}_S| = 2^\omega$.

The resulting structure of all arbitrary ω-sequences is depicted in the form of the $2^\omega \times 2^\omega$ matrix $\mathcal{M}$ in Figure 6.1.

In this matrix $\mathcal{M}$, the columns enumerate all the specific ω-sequences

$$o_1, o_2, o_3, \ldots, o_\alpha, \ldots$$

Figure 6.1 Generic ω-sequences.

The rows enumerate the states in the configuration space $\mathcal{C}_S$. A thread through $\mathcal{M}$ (the boxed entries are elements of one such thread) is an arbitrary ω-sequence σ, where

$$\langle o_\alpha, \mathcal{C}_\beta \rangle \in \sigma =: \text{ the value of } \sigma \text{ at } \mathcal{C}_\beta \text{ is } o_\alpha.$$

Again Figure 6.1 contains some excess structure because the configuration space is not *intrinsically* ordered in any particular way.

The notion of being κ-generic (Section 4.1) can be lifted from arbitrary numbers to generic ω-sequences in the following way:

Definition 6.1. A generic ω-sequence σ is κ-generic if for each specific ω-sequence o, there are κ many states where σ takes the value o.

Generic ω-sequences can be taken to consist of specific and arbitrary natural numbers. Both the specific and the arbitrary numbers are arbitrary A's, i.e. objects that in each possible world is in the state of being a specific object, i.e. one of $a_0, a_1, a_2, \ldots$ There are 2^{2^ω} such; each of them is a number belonging to each generic ω-sequence.

Let f be any generic ω-sequence, and recall that every state is a specific ω-sequence. The specific number 0 of f, denoted as 0_f, is the arbitrary object a such that for every possible world w, if f is in state s at w, then a at w is in the state of being the initial element of s. Similarly, 1_f is the arbitrary object a such that for every possible world w, if f is in state s at w, then a at w is in the state of being the second element of s, and so on and so forth. An arbitrary natural number of f is an arbitrary A that is not n_f for any $n \in \mathbb{N}$.

This entails that whether an arbitrary A is a specific number, and if so, which one, is very much a generic ω-sequence-specific matter: any arbitrary A is 0_f for many generic ω-sequences f, and an arbitrary number for many other generic ω-sequences.

Elementary functions on arbitrary numbers of generic ω-sequences are defined point-wise on the worlds in the obvious way, and they are of course also ω-sequence-relative. For instance, the sum $a +_f b$ is such that for every possible world w, if f takes the value s at w, a is in the state of being the kth element of s at w, and b is in the state of being the lth element of s at w, then $a +_f b$ takes the $k + l$th element of s at w.

Elementary operations can then straightforwardly be seen to satisfy the familiar properties of arithmetical operations (such as commutativity of $+_f$, for instance). However, the (specific and arbitrary) numbers of generic ω-sequences are not well-ordered in any obvious way, so it is not clear at this stage how a general principle of induction, covering both specific and arbitrary numbers of f, can be formulated.

We started this section with the assumption that there are countably many objects $a_0, a_1, a_2, \ldots$ What if there are more than countably many objects? Then the argument developed in this section still goes through: there will be many generic ω-sequences. But the calculations of how many there are, and how many arbitrary natural numbers there are, then have to be revised.

In fact, I have (in the interest of perspicuousness) not been quite correct in this section in taking into consideration only the objects there actually are. It should be clear that as long as there *could have been* infinitely many objects, the reasoning in this section applies, and there are (many) generic ω-sequences. What if there could *not* have been infinitely many objects? Then there are not and could not have been such things as generic ω-sequences.

6.4 The Generic Natural Number Structure

Consider again all the objects there are. And again suppose that there is a countable infinity of them. Consider all ways s_α of ordering all or some of them into an ω-sequence. There will again be 2^ω such ways.

Now suppose that, in the spirit of Benacerraf, we take a structuralist stance to arithmetic: 'any ω-sequence will do' (Section 5.4.1). Then take the entity **N** whose nature is that it *can* be (and can only be) any of those ω-sequences. What this corresponds to, of course, is to take *one* diagonal ω-sequence (or 1-generic ω-sequence) in the system of arbitrary ω-sequences, for instance the boxed arbitrary ω-sequence in Figure 6.2. Call this entity *the full generic natural number structure*.

As before, we can describe what it means to be a specific or arbitrary number (and we can define elementary operations on these numbers in

$$\begin{array}{ccccc}
\boxed{\langle o_1, C_1\rangle} & \langle o_2, C_1\rangle & \ldots & \langle o_\alpha, C_1\rangle & \ldots \\
\langle o_1, C_2\rangle & \boxed{\langle o_2, C_2\rangle} & \ldots & \langle o_\alpha, C_2\rangle & \ldots \\
\vdots & \vdots & \vdots & \vdots & \vdots \\
\langle o_1, C_\alpha\rangle & \langle o_2, C_\alpha\rangle & \ldots & \boxed{\langle o_\alpha, C_\alpha\rangle} & \ldots \\
\vdots & \vdots & \vdots & \vdots & \vdots
\end{array}$$

Figure 6.2 The diagonal ω-sequence **N**.

natural ways). Formally, this means taking an arbitrary number to be a sequence of natural numbers of length 2^ω. This is different from the way we modelled arbitrary natural numbers in Section 4.1, outside the structure of mathematical structuralism: there we modelled arbitrary numbers as sequences of natural numbers of length ω. The reason, of course, is that the mathematical structuralist recognises more states.

The generic system **N** is significantly different from generic system depicted in Figure 4.1 of Section 4.1. For instance, we have seen that the generic system of Section 4.1 contains (many) diagonal numbers; for elementary cardinality reasons, the generic system **N** does not contain any diagonal numbers. In sum, adopting a particular structuralist viewpoint may make a real difference for the theory of arbitrary natural numbers.

In a structuralist spirit, the thesis can be formulated that **N** is the structure that arithmetic is about. We will discuss this thesis in some philosophical detail in Chapter 8. Here I pause to observe that there appears to be an immediate conflict with what was said in the previous section. There it was maintained that there are *many* (full) generic ω-sequences. But I just talked about *the* full generic natural number structure **N**. What gives?

The apparent conflict results from a difference between regarding **N** as a structure or universe on the one hand, and regarding **N** as an element in a larger structure or universe on the other hand. Consider any other full generic ω-sequence **M** belonging to the universe of generic ω-sequences. Now consider this ω-sequence **M** on its own rather than as an element of the system of arbitrary ω-sequences. Then, given our ontological assumption that we 'collapse' or identify states that are identical, what we are left with is just **N**.

So if you are doing number theory (without making use of 'higher mathematics' to obtain number theoretic results), then you are working within *the* generic natural number structure. But the generic natural number structure is itself an entity belonging to a larger universe. If you are making use of generic ω-sequences or are investigating them as as a class, then you are dealing with multiple copies of the generic natural number structure that are modally correlated with each other.

We *could* at this point go on to construct a model of arithmetic on the basis of the states in **N** by means of an ultrapower construction (Bell and Slomson 2006). We could consider a filter on the collection of the ω-sequences s_α, now considered as an index set. This filter could then be extended to some ultrafilter $\mathcal{U}$, and then we could take the ultrapower

$$\prod_{\mathcal{U}} s_\alpha.$$

This would then be a model of arithmetic. If $\mathcal{U}$ is a principal ultrafilter, it would be isomorphic to the standard model; if $\mathcal{U}$ is non-principal, this model would be non-standard. But this is *not* what I am interested in. Instead, I am interested in **N** itself as a generic system.

This is an important point. We have seen in Section 4.6 how Fine believes that the connection between arbitrary object theory and the theory of infinitesimals should be pursued. But I believe that the genius of what Fine has done is to stop half-way into the process of building a non-standard model (i.e. *before* introducing a filter on the index set), and investigate the 'half-finished' product that has been constructed at that stage as a structure that is of considerable metaphysical significance in and of itself.

6.5 Other Arbitrary Systems

This account latches on to how mathematicians treat structures. Recall the example of the complete three-element graph of Section 5.5.1. Mathematicians speak of *the* complete two-element graph as a structure. But they also say that the complete three-element graph contains three complete two-element graphs as subgraphs. This again means that taken by itself, there is only one complete two-element graph. But the complete two-element graph can also function as an object in a larger structure, and then its components are coordinated with other elements of this larger structure in multiple ways.

The account that is developed in the previous section and in the present section can be generalised to any individual mathematical structure. Suppose we want to know what a particular mathematical structure S (of size κ) is, from the perspective of arbitrary object theory. Then if we start with a collection D with $|D| = \kappa$, there will be a collection C_D of isomorphic systems that instantiate S and are based on D. This collection C_D then serves as the state space for the system $\mathbf{A}_S$ of arbitrary S-es. This system will contain 1-generic arbitrary S-es. Any such 1-generic arbitrary S can be taken to be the 'perspective from within': it can be identified with the mathematical structure S.

Mathematical structuralism is not wedded to the categoricity of arithmetic or of any other mathematical discipline. We have seen how Resnik voices strong reservations about second-order logic (see Section 5.4.1). Yet he is one of the leading defenders of mathematical structuralism. Conceiving of mathematical structures as arbitrary systems also does not automatically commit you to taking a discipline such as arithmetic to be categorical.

A 'first-orderist' conception of the natural numbers structure as an arbitrary system would go along the following lines. As before, start with a countable collection A of objects. But now let the states of the natural number structure not be all ω-sequences, but rather all non-standard models (plus the standard model) $\langle A, 0, +, \times \rangle$ of first-order Peano Arithmetic (or some other recursively axiomatisable first-order arithmetical theory, or even true first-order arithmetic). The state space Ω of the (first-orderist) arbitrary natural number system consists of all those models – there are 2^{ω} such.

Then, as before, let the arbitrary numbers of this arbitrary system consists of functions from Ω to elements of A. The specific natural numbers are canonically embedded into this arbitrary system. Addition and multiplication of arbitrary numbers can again be defined pointwise. Let us call the resulting arbitrary system $\mathbf{N}^*$. Since not all states of $\mathbf{N}^*$ are isomorphic, $\mathbf{N}^*$ is, in the terminology of Section 6.1, an arbitrary system but not a generic system.

6.6 The Computable Generic ω-Sequence

Suppose that the generic natural number structure $\mathbf{N}$ is generated, along the lines described in the previous section, from a countable plurality of *physical* objects. Then we have no guarantee even that the successor relation of *any* of the ω-sequences on A is a bona fide physical relation (such as height, or weight). It is also not clear that the successor relation on any of these ω-sequences is practically computable, even when we make the usual idealisations such as unlimited memory, no hardware failures, etcetera.

It seems virtually certain that not *all* of the 2^{ω} ω-sequences of $\mathbf{N}$ are practically computable: $\mathbf{N}$ surely contains ω-sequences that we cannot compute on. You might want to disqualify such ω-sequences from being possible states of the generic ω-sequence. You might want to insist that $\mathbf{N}$ can only ever be in the state of being a *computable* ω-sequence. (Of course this viewpoint is not forced upon you.)

This is equivalent to saying that the generic natural number structure should consist of *computable notation systems*, as defined in Section 5.6. So a computable notation system for the natural numbers is a collection $E \subseteq A$ of 'expressions' (physical entities) together with a practically computable successor relation on E that well-orders E with order type ω. (If the successor relation is computable, then elementary functions on the notation system are also computable.) These computable notation systems form a practically computable isomorphism type (Halbach and Horsten 2006). So we may (and often will) without loss of generality take them to be generated by recursively permuting Hilbert's stroke notation system. So let us define the *computable generic natural number structure* $\mathbf{N}_C$ as the entity that can be (and can only be) in the state of being any such computable notation system. This structure is of some significance both for epistemological reasons and for technical reasons.

We have seen in Section 4.4 how Parsons has argued that the stroke notation system is significant from an epistemological point of view because we can acquire intuitive knowledge about it (Parsons 2008, Chapter 7). For instance, the successor function is given by concatenation with one stroke; the principle that any string has a 'successor' is intuitive (Parsons 2008, Section 29). Also, other elementary 'arithmetical' principles governing strings are intuitive (Parsons 2008, Section 41). It is not clear at all that on Parsons' conception of what it means to be intuitive, all notation systems inter-translatable with stroke notation are also intuitive, but that need not concern us at present.

The notation systems in $\mathbf{N}_C$ can be taken to be generated from the stroke notation system by recursive permutations. Therefore $\mathbf{N}_C$ has ω states. This makes it from a formal point of view much more similar to the way in which arbitrary numbers were modelled in Section 4.1 than to the generic system $\mathbf{N}$ of the previous section. We will see in Chapter 8 that the structure $\mathbf{N}_C$ is from a logical point of view more tractable than the full generic natural number structure $\mathbf{N}$.

6.7 The Arbitrary Countable Graph

Let us now consider *algebraic* theories from the perspective of the theory of generic systems. As an example, take the theory of countable unlabelled simple graphs. By analogy with the full generic ω-sequence and the space of full generic ω-sequences, we can define the arbitrary countable simple graph and the space of arbitrary countable simple graphs.

Suppose we are given a countable collection $V = \{v_0, v_1, v_2, \ldots\}$ of objects that will figure in the vertex sets of the graphs. We consider all possible (specific) countable simple graphs on subsets of V. These are *systems* that have as their countably many vertices some of the V's; their vertices are related to each other by a relation which is symmetric and anti-reflexive. There are 2^ω such countable simple graphs.

The (fully) arbitrary countable simple graph **G** is the entity that can be any of these countable simple graphs and can be in no other states. It is an arbitrary system but not a generic system, for not all of its states are isomorphic.

The arbitrary system **G** can be taken to have vertices and edges. A vertex of **G** is an arbitrary object that in any possible situation is a vertex of the system that **G** is in that possible situation. An edge of **G** is an arbitrary object that in any possible situation is an edge of the system that **G** is in that possible situation. Thus the vertices and edges of **G** are arbitrary entities.

From a higher-order point of view, there are many arbitrary countable simple graphs, some more arbitrary than others. Each of them contains vertices and edges in the same way that **G** does. A configuration space is associated with the arbitrary countable simple graphs. The arbitrary countable simple graphs can be taken to 'live' in this space: each point can be seen as a *possible situation* in which all arbitrary countable simple graphs take on a specific value. If we disallow pairs of arbitrary countable simple graphs that have the same modal profile (i.e. are exactly the same systems in exactly the same possible worlds), then there are $2^{2^{\omega}}$ arbitrary countable simple graphs.

Particular countable simple graphs are represented in the space of arbitrary countable simple graphs. A particular graph A is represented by a generic graph $\mathbf{G}_A$ that is in every possible situation isomorphic with A. Consider for instance the complete two-element graph. This is represented by a graph $\mathbf{G}_2$ that can be all and only the systems that have the structure of the complete two-element graph. Note that this representation is not unique: there are many generic graphs in the space of arbitrary countable simple graphs that represent $\mathbf{G}_2$.

A celebrated theorem from countable graph theory by Erdős and Rényi (1963) says that there exists a simple graph **R** with the following property. If a countable graph is chosen at random, by selecting edges independently with probability $\frac{1}{2}$ from the set of two-element subsets of the vertex set, then almost surely (i.e. with probability 1), the resulting graph is isomorphic to **R**. This graph **R** is called the *Radó graph*. In the present context, this theorem can be taken to say that with probability 1, the arbitrary countable graph **G** is the Radó graph (Cameron 1997). Thus Erdős and Rényi's theorem can be seen as a *probabilistic categoricity theorem* for countable graph theory.

We could now do the same for some other algebraic theory, such as countable group theory for instance. We could also move from countable

Figure 6.3 The graph $\mathbf{G}_{a-b,c}$.

mathematics to uncountable mathematics, by considering e.g. the theory of
the real numbers. But the general procedure is clear, so there is no need to
go into details.

In his doctoral dissertation, Lawvere set out to specify a *category* (in the
sense of Category Theory) that is a generic group (or a generic ring, or a
generic field) (Lawvere 1963). And that category would be defined in such a
way that then any specific group (or ring, or field) can be identified with
a functor on that group (or ring, or field) to the category of sets. Such
categories are called *Lawvere theories*.

It might be interesting to see how Lawvere's approach relates to my
attempt to define 'the arbitrary countable graph'.[1] One general advantage
of category theoretic representations is that they contain less 'redundancy'
than set theoretic representations typically do. Recall, for instance, how $\mathbf{G}_2$
is represented in many different ways in the space of arbitrary countable
simple graphs. Lawvere theories avoid this kind of redundancy. On the
other hand, Lawvere theories are categorical representations of *equational
theories*. The approach that was sketched in this section is more general in
the sense that it can be applied to *all* algebraic theories, including those that
do not have an equational representation.

6.8 Individual Structures

In Section 6.4 we saw how the natural number structure can be seen as
a generic system. The natural number structure is special because it does
not admit non-trivial automorphisms. I will now briefly look at individual
structures that do admit non-trivial automorphisms.

As an illustrative example, consider the simple unlabelled graph $\mathbf{G}_{a-b,c}$
in Figure 6.3, which consists of three vertices of which exactly one is unconnected.

---

[1] Thanks to Richard Pettigrew for making this suggestion.

Figure 6.4 The systems S_1, S_2, S_3.

It has a non-trivial automorphism that swaps its two edge-connected vertices. How can we see this structure as a generic object?

The systems that instantiate $\mathbf{G}_{a-b,c}$ contain three objects. So let us, without loss of generality, suppose that these systems are all built from three concrete objects: a, b, c. From these objects, three systems S_1, S_2, S_3 that instantiate $\mathbf{G}_{a-b,c}$ can be built. These systems are depicted in Figure 6.4.

As before, these three systems S_1, S_2, S_3 can be seen as forming a state space. Threads through this state space are arbitrary objects. How can we find canonical representations of the three vertices of $\mathbf{G}_{a-b,c}$?

The thread t_c which in each state picks out the unconnected object is of course the only good candidate for 'being' the unconnected vertex of $\mathbf{G}_{a-b,c}$, which is a *particular* vertex. Moreover, there are threads that play the role of the unconnected vertex in one state, and the role of one of the connected vertices of $\mathbf{G}_{a-b,c}$ in another: those are very *arbitrary* vertices. The problem is that even for the small graph $\mathbf{G}_{a-b,c}$, there are many – eight, to be precise –

threads that only ever play the role of one of the connected vertices. Let us call these threads *potential connected vertices*. Since $\mathbf{G}_{a-b,c}$ only contains *two* connected vertices, the potential connected vertices cannot all be one of them. What to do?

The potential connected vertices are correlated with each other. Given a choice of one potential connected vertex, there is only one possible choice for the other. There are four such pairs:

$$\{t_{a_1}, t_{b_1}\}, \{t_{a_2}, t_{b_2}\}, \{t_{a_3}, t_{b_3}\}, \{t_{a_4}, t_{b_4}\}.$$

So there are four candidates $\{t_{a_i}, t_{b_i}, c\}_{i \leq 4}$ for 'being' the vertices of the graph $\mathbf{G}_{a-b,c}$. This implies that we have to move one level up in degree of abstraction. The pair of connected vertices of $\mathbf{G}_{a-b,c}$ is a system consisting of two *second-order arbitrary objects* t_1^2 and t_2^2. This system can be in one of four states. But they are coordinated: if t_1^2 is in state t_{a_i}, then t_2^2 is in state t_{b_i}.

In a similar way we can 'lift' the edge relation to t_1^2 and t_2^2 from the states in which the arbitrary objects t_{a_i} and t_{b_i} can be. All of this is just modelling. Therefore, as in the case of the generic natural number system, none of this means that the graph $\mathbf{G}_{a-b,c}$ *is* the (particular) graph that has t_1^2, t_2^2, t_c as elements and $\langle t_1^2, t_2^2 \rangle$ as edge relation.

One worry remains. We saw that there are eight potential connected vertices. Should then not each of t_1^2 and t_2^2 be able to be in each of these eight states? The answer is no. In a state where one of t_1^2 and t_2^2 is t_{a_i} and the other is t_{b_i}, there is simply no matter of fact whether t_1^2 is t_{a_i} and t_2^2 is t_{b_i}, or t_1^2 is t_{b_i} and t_2^2 is t_{a_i}. It would be a mistake to think that t_1^2 leads an existence independent of t_2^2, and that it makes sense to ask about given state: is t_1^2 identical to t_{b_i} and t_2^2 to t_{a_i} in this state, or is it the other way round?

I have modelled the vertices and edges of $\mathbf{G}_{a-b,c}$ as *second-order* arbitrary objects. They were taken to be arbitrary objects that can be in states of being some arbitrary object that itself can be in states of being a specific object. Natural numbers, on the other hand, were in Section 6.4 modelled as first-order arbitrary objects.

The reason for this can be seen by inspection of the graph $\mathbf{G}_{a-b,c}$ in Figure 6.3. The graph $\mathbf{G}_{a-b,c}$ admits of exactly one *non-trivial automorphism* – let us call it f. The unconnected vertex is the only element of the graph $\mathbf{G}_{a-b,c}$ that is not moved by f. The unconnected vertex was perfectly represented by t_c. It was only because of the connected vertices, which are moved by f, that we were forced to move up one level of abstraction.

At this point we see that this way of modelling individual structures as arbitrary entities that can be identified with or based on generic systems, generalises. For instance, we could now explain how the cyclic group of order 4 ($\mathbf{C_4}$), which also admits of a non-trivial automorphism, can be seen as an arbitrary entity. Whenever a mathematical structure admits nontrivial automorphisms, a second-order arbitrary object is needed to represent it faithfully. In this way, there is a sense in which the graph $\mathbf{G}_{a-b,c}$ or the group $\mathbf{C_4}$ is *more abstract* than the natural number structure.

In this context, we may ask:

Question 6.2. Is *every* algebraic mathematical discipline concerned with at least some individual mathematical structures that admit non-trivial automorphisms?

A positive answer to this question would plausibly exclude set theory from the class of algebraic disciplines. In any case, it is evident that not every categorical sub-field of mathematics is concerned with a rigid structure. As we have seen on page in Section 5.5, complex analysis is a counterexample.

6.9 The Generic Hierarchy

Nothing has yet been said about the nature of the objects of the underlying domain of generic systems (ω-sequences, graphs, groups): the choice of the underlying plurality of objects has so far played the role of an *external parameter* of the model. But the question of the nature of the objects of the underlying domain is one that cannot altogether be avoided.

Nothing in mathematical practice dictates that the underlying objects should be physical in nature. But clearly *many* of them are required if we want to extend the view described in this chapter beyond countable mathematics.

The *universe of sets* contains an ample supply of objects for our mathematical objects and structures. It might seem *ad hoc* if the subject matter of set theory would fall altogether outside the scope of the theory of arbitrary systems.

If we admit higher-order arbitrary objects, then the theory of arbitrary objects provides the resources for giving a natural explication of what an iterative hierarchy of set theoretic ranks may be taken to consist of. In the light of Definition 4.6, a metaphysical account of how this might work goes along the following lines:

- **Stage 0**
  Take some specific object o to be given.

- **Stage 1**

  Consider the arbitrary object which can only be in the state of being the specific object o; let us denote this arbitrary object as $\langle o \rangle$. Observe that $o \neq \langle o \rangle$: unlike an arbitrary object, a specific object is not the sort of entity that can be in states.

- **Stage 2**

  (2a) Consider the higher-order arbitrary object that can only be in the state of being the arbitrary object $\langle o \rangle$. Denote this higher-order arbitrary object as $\langle \langle o \rangle \rangle$.

  (2b) Consider an higher-order arbitrary object that can only be in one of the following two states: being the specific object o, or being the arbitrary object $\langle o \rangle$. Denote this object as $\langle o, \langle o \rangle \rangle$.

- **Stage ω**

  Collect the arbitrary objects that have been generated in the finite stages.

- **Later stages**

  Continue in this way into the transfinite.

In this way, a hierarchy of higher-order arbitrary objects, based on a specific object, is built up. The simple idea of course is to view arbitrary objects as (non-empty) sets, and their states as their elements. This leaves the specific object on which the hierarchy is based, which is not the sort of thing that can be in states, playing the role of the empty set. So we read the pointy brackets $\langle \ldots \rangle$ as curly brackets $\{ \ldots \}$, and o as $\emptyset$. For example, the set $\{\emptyset, \{\emptyset\}\}$ is the arbitrary object which can be specific object o, but which can also be the arbitrary object which can only be the specific object o.

Let us call this the *generic hierarchy*. The concept that it might *perhaps* be taken to capture is the *combinatorial* set concept, which takes a set of Xs to be the result of an arbitrary selection process.

In this generic hierarchy, the empty set is a specific object, whereas all non-empty sets are arbitrary objects. This might be considered an unnatural aspect of the account. But it can be avoided if we admit, as we have contemplated doing in Section 3.8.2, arbitrary objects that are *undefined* in some states. On the picture that I have presented, possibly being in the state of being x corresponds to having x as a member. It would therefore be natural to take the empty set to be the (unique) arbitrary object that is everywhere undefined, i.e. that does not have a value in any state. The further stages of the generic hierarchy can then be defined from this starting point in the same way as above. This would result in a 'pure' generic hierarchy rather than one that is based on one *Urelement*.

We have seen how the distinction between a and $\langle a \rangle$ corresponds to the distinction between a set and its singleton. It is not clear whether this metaphysical account gives an adequate explanation of what Lewis (1991) has dubbed the 'mystery of the singleton relation'. Some may worry that the distinction between a specific object and the arbitrary object that can only ever be in the state of being that specific object is no better motivated than the relation between an object and its singleton.

An important observation is that the generic hierarchy is the result of an extreme *flattening* or extensionalisation of what is at bottom a much more intensional notion of set. There are, for example, in fact *two* higher-order arbitrary objects that can play the role of the set $\{\emptyset, \{\emptyset\}\}$ in stage (2b). These two arbitrary objects are anti-correlated; if one of them is in the state of being o, then the other is in the state of being $\langle o \rangle$ and vice versa. (So one might denote these two objects as $\langle o, \langle o \rangle \rangle$ and $\langle \langle o \rangle, o \rangle$, respectively.)

The engine of the iterative hierarchy of sets is the full power set operation. This operation drops out naturally in the framework of the theory of arbitrary objects: the description of stages shows how at every stage $\alpha + 1$, there are 2^α arbitrary objects that naturally represent all the subsets of the domain of sets that exist at stage α.[2]

All this is a *non-structuralist* interpretation of set theory. But I have in Section 5.7 endorsed Burgess' arguments for the thesis that with respect to matters of mathematical structure, set theory occupies a special and perhaps even unique position among the mathematical disciplines.

---

[2] I do not claim that the present account has much new to say about the motivation of some of the other powerful axioms of standard set theory such as Infinity or Replacement.

7 | Kit Fine

The idea to relate the notion of arbitrary object to that of an *ante rem* mathematical structure has occurred to different philosophers at different times. But as far as I know, this idea has not been worked out in great detail.

In his review of Parsons' *Mathematical Thought and Its Objects*, Burgess (2008, p. 403) writes:

Though not usually presented in this way, [eliminative structuralism and non-eliminative structuralism about the real numbers] may be taken to start with the idea that 'the real numbers' means 'the arbitrary complete ordered field', and then diverge over the interpretation of 'arbitrary' ... On the minority view, as in Fine (1985b) the arbitrary F is a specific F but an extraordinary one in that it has no properties not shared by all Fs (though it is distinguished from ordinary Fs by the 'meta'-property' of being arbitrary).

But Burgess then quickly moves on from this initial characterisation of non-eliminative structuralism in terms of arbitrary objects to Shapiro's version of non-eliminative structuralism.

Ten years earlier, in Section VI of Fine (1998), Fine sketches how a philosophical account of mathematical structure can be formulated within the framework of the theory of arbitrary objects. Fine does not work out his proposal in detail, but it is clear that his proposal differs from the one that I have sketched in the previous chapters in certain respects. These differences are in part explained by a disagreement between Fine and me about how the theory of arbitrary objects should be articulated in the first place.

In this chapter I start by giving an exposition of Fine's theory of arbitrary objects. In doing so, I will assume – since I have not seen any evidence to the contrary – that Fine has not fundamentally changed his view about the nature and properties of arbitrary objects from his earliest publication on the subject (1983) to his most recent one (2018). Towards the end of the chapter I explain how Fine deploys his theory of arbitrary objects in applications to the philosophy of mathematics.

I will isolate open metaphysical questions, but I refrain in this chapter from raising objections to Fine's position, except on relatively minor issues. One of the main messages will be that Fine's metaphysical account of arbitrary objects remains hitherto in a state of incomplete development. The metaphysical benefits of completing Fine's picture of arbitrary objects will surely be considerable. This does not mean that I am in agreement with Fine on all fundamental aspects of the theory of arbitrary objects and their applications. But I largely postpone a critical evaluation of the differences between Fine's theory of mathematical structures and the one that I am proposing until the next chapter.

7.1 Concept and Object

In his *Principles of Mathematics*, Russell (1903b, p. 42) took the grammar of ordinary language as a guide to what exists and how it exists. This led him to drawing a fundamental distinction between concepts and things. It led him to taking an expression of the form 'any *a*' to express a concept that, in the context of a proposition, also has a denotation. We have seen that Russell was somewhat unclear about the denotation of this concept (see Section 3.2). He described the denotation as being 'one *variable* object *a*', but also at least suggested that the notation was a 'variable conjunction' of some sort.

Russell then soon moved away from taking the concept associated with an expression of the form 'any *a*' to have a denotation at all in the context of a proposition of the form 'any *a* is *F*'. At the time when he is writing *On Denoting*, Russell sees no need anymore for a 'metaphysical' correlate of the concept expressed by the expression 'any *a*'.

Like Russell's variable objects, Carnap's individual concepts bear a resemblance to Fine's arbitrary objects. Carnap (1956) introduced individual concepts to describe meanings of expressions that can occupy subject position in a sentence. Officially, they are functions from state descriptions to objects. From a methodological point of view, individual concepts played a different role for Carnap than variable objects played for Russell. In the period when he wrote *Principles of Mathematics*, Russell used the method of naive metaphysics – even though he was more guided by natural language than a naive metaphysician is likely to be. But Carnap used a very different methodology: he introduced individual concepts as *theoretical entities* that are useful in semantics.

Carnap was fickle about the extent to which individual concepts (and other abstract entities such as mathematical objects) should from an ontological point of view be taken seriously. He wanted them to do semantic work (Carnap 1950). But in his formal framework he did not allow the formation of real identities between them (see Chapter 9). Indeed, this troubled Quine, who thought that you should have a criterion of identity for every kind of entities that you allow in your ontology. Carnap could reply that he *was* able to offer a criterion of identity for individual concepts:

$$\text{individual concept } a = \text{individual concept } b \Leftrightarrow$$
$$a(w) = b(w) \text{ for every state description } w.$$

But this would not satisfy Quine. He would point out that *such* individual concepts are too course grained for adequately representing meanings of terms. Moreover, and more importantly for Quine (1948), the proposed criterion immediately invites a further question: what is a correct and informative criterion of identity for state descriptions or possible worlds more generally?

The ensuing debate has filled thousands of pages, and there is fortunately no need for rehearsing the main tenets of this philosophical discussion here. What is somewhat relevant for us is that Carnap and Quine were *not* doing naive metaphysics. In this respect, Fine's theory of arbitrary objects is much closer in spirit to Russell's early theory of variables than to Carnap's theory of individual concepts.

In his theory of arbitrary objects, Fine takes a much more resolutely metaphysical stance than Carnap. Moreover, he is much clearer and more explicit about what arbitrary objects are than Russell was when he still believed that there were such entities. Fine (1985b, p. 6) begins his book on arbitrary objects as follows:

There is the following view. In addition to individual objects, there are arbitrary objects: in addition to individual numbers, arbitrary numbers; in addition to individual men, arbitrary men. With each arbitrary object is associated an appropriate range of individual objects, its values: with each arbitrary number, the range of individual numbers; with each arbitrary man, the range of individual men. An arbitrary object has those properties common to the individual objects in its range. So an arbitrary number is odd or even, an arbitrary man is mortal, since each individual number is odd or even, each individual man is mortal. On the other hand, an arbitrary number fails to be prime, an arbitrary man fails to be a philosopher, since some individual number is not prime, some individual man is not a philosopher.

Such a view used to be quite common, but has now fallen into complete disrepute.

Despite its uncommonnness, this is pretty much the view that Fine seems to hold in his book. But when pushed on the question of realism, Fine shuffles his feet, much like the average mathematician who is a platonist on weekdays but retreats to a non-platonist view when pushed by a philosopher. In his official view about the realism question concerning arbitrary objects, Fine (1985b, p. 7) expresses Carnapian reservations:

If now I am asked whether there are arbitrary objects, I will answer according to the intended use of 'there are'. If it is the ontological significant sense, then I am happy to agree with my opponent and say 'no'. I have a sufficiently robust sense of reality not to want to people my world with arbitrary numbers and arbitrary men. Indeed, I may be sufficiently robust not even to want individual numbers or individual men in my world. But if the intended sense is ontologically neutral, then my answer is a decided 'yes'. I have, it seems to me, as much reason to affirm that there are arbitrary numbers in this sense as the nominalist has to affirm that there are numbers.

This seems to imply a reductionist programme of some sort – a programme that, as far as I know, is not worked out in the book or in other writings by Fine. Given his more recent writings on metaphysical methodology, I doubt that Fine would today answer the question of realism about arbitrary objects in exactly this way. Be that as it may: I am sceptical that arbitrary objects can metaphysically be reduced to a nominalistically acceptable basis.

Nonetheless, what is more important than all this is that Fine's metaphysical stance regarding arbitrary objects brings about a shift. Carnap's individual concepts, and to some extent also early Russell's variables, are not much more than reified predicates. But Fine resolutely takes arbitrary objects to be *objects*. He is very much concerned with relations of numerical identity and difference, and dependence between arbitrary objects. And the relation between arbitrary objects of coinciding in a possible state comes to occupy centre stage.

If we furthermore develop the theory of arbitrary objects in the way that I have done in the previous chapters, then there are simply too many arbitrary objects for there to be any hope of ontologically reducing them to the specific objects that serve as their value ranges. But, as we will now see, Fine develops the theory of arbitrary objects in a somewhat different way.

7.2 Variable Objects and Dependence

Fine starts by taking the idea of arbitrary or *variable* objects ontologically seriously. (In an attempt to avoid confusion, I will speak from now on mostly of variable objects when discussing Fine's conception, and continue to talk of arbitrary objects to refer to my conception.) Variable objects are objects that, by virtue of their intrinsic nature, take *values* (Fine 1998, p. 609). This means that quantification over variable objects comes with a *range of values* for the variable objects. Such quantification can be restricted by means of predicates. This allows us then also to quantify over variable objects with a restricted range of values.

It is not completely clear to me what Fine takes the metaphysical nature of having a value to be. For now, let us understand taking a value as being in a state, as that phrase has been used in earlier chapters. We will see Section 7.4 how this reading will certainly have to be qualified somewhat if we want to represent Fine's account accurately.

On Fine's (1983, p. 68) conception, variable objects can depend on other variable objects. Fine (1998, p. 609) distinguishes two kinds of dependence between arbitrary objects in this context: *one-way dependence*, and two-way dependence or *co-dependence*. Moreover, there are variable objects that do not depend on other variable objects: these are said to be *independent* (p. 610).

The notions of dependence and independence can be illustrated as follows. Consider the following two fragments of mathematical discourse (Fine 1998, p. 611):

Example 7.1.
LET x and y be such that: x and y are reals and $y = -x$.

Example 7.2.
LET x' be such that: x' is real.
LET y' be such that: $y' = -x'$.

The variable objects x and y in example 7.1 are co-dependent. The variable object y' in example 7.2 one-way depends on the variable object x'. Moreover, x' is an independent variable object.[1]

Fine does not tell us what *makes* it the case that there is two-way rather than one-way dependence between x and y in the case of example 7.1 and

---

[1] I am actually a bit sloppy here by identifying the variable signs in the examples with the variable objects 'assigned' to them (Fine 1998, p. 610).

only one-way dependence between x' and y' in example 7.2. Of course the linguistic differences in the two discourse fragments can be of no help in answering these questions. All they do, if they succeed in what they aim to do – and how do they do it? – is to latch on to a metaphysical distinction that is already there. After all, Fine (1998, p. 609) says:

I wish to think of [variable objects], however, as a certain kind of object ... They will not be linguistic in character. They will not be conventional symbols for their values, but abstract objects which assume those values by way of their intrinsic nature.

Fine (1985b, pp. 24–25) has argued that some conditions, such as transitivity and well-foundedness, can reasonably be taken to hold of this dependence relation between variable objects. If well-foundedness is in place, then a notion of *rank* can be associated with variable objects in the familiar way.

The requirement of transitivity requires no comment: it is unobjectionable. Well-foundedness (or Foundation) is a very substantial claim, and Fine has always been keenly aware of that. It entails that the dependence relation is irreflexive. This consequence seems unproblematic: the issue of the reflexiveness of the dependence relation is somewhat of a non-question. Either we take it to be reflexive ('every variable object depends on itself') or we say that the relation is anti-reflexive ('no variable object depends on itself'): it is hard to give more than a conventional meaning to the decision what to say here. It does not, however, on the face of it seem to make sense to say that some variable objects depend on themselves whilst others don't. Well-foundedness also implies that dependence is anti-symmetric. So it entails that the dependence relation is a *partial ordering relation*.

Anti-symmetry is a debatable restriction on the dependence relation. Indeed, our discussion of example 7.1 above shows that Fine must have abandoned the well-foundedness restriction in his later work.

Even in his earlier work, Fine only gives a somewhat guarded argument for Foundation (Fine 1985b, p. 25):

Foundation is most reasonable when the dependency relation $a \prec b$ is specifically construed as meaning that the value of b can be determined on the basis of the value of a (and perhaps the value of some other [variable objects] as well). Foundation then says that the attempt to determine the value of [a variable object] on the basis of other [variable objects] will eventually lead to an end, that will terminate in [variable objects] whose values are just given.

In this passage, an *epistemic* reading of the word 'determination' is presumably not intended: if it were, then the argument would of course be unconvincing. So let us assume that a metaphysical interpretation of determination is intended. But then also it is not clear why we should be swayed by this argument. It sounds much like the Aristotelian argument for the existence of an *Unmoved Mover*. In sum, it seems right for Fine to have dropped the well-foundedness constraint on the dependence relation in his later work.[2]

We will see in Section 7.3.1 how, theoretically, the partial ordering of variable objects associated with a set-sized value range F of specific objects is proper class sized on Fine's (1983, p. 71) account. But in practice, in a given context, for a given F, we are dealing with a (smallish) *finite* partial ordering of arbitrary F's: each of them is introduced into the conversation, starting with the root of the partial ordering (the independent F), and working upwards in stages along the ranks of the dependent arbitrary F's.

A *system* of variable objects is a dependence relation-*connected* plurality of variable objects that is *closed* under the dependence relation (Fine 1998, pp. 609–610).

7.3 Identity, Systems, and Comprehension

In Fine's theory of variable objects, as in mine, identity and comprehension principles play a central role. But we will see that Fine does not to seem to have a firmly settled view on some of the issues involved.

7.3.1 Identity. Let us start with identity. In his first publication on arbitrary objects, Fine (1983, p. 69) wrote, without referring explicitly to *systems* of variable objects:[3]

Suppose first that a and b are independent [variable] objects. Then we say that a = b iff their ranges are the same.

This leads him to answer Frege's question how many arbitrary numbers there are (see Section 4.1) along the following lines (Fine 1983, p. 71):

We may sink this into the more general question: what arbitrary objects are there? Consider first the arbitrary objects that take their values from a given

---

[2] Fine already mentions in Fine (1985b, p. 28) that for some purposes well-foundedness should be dropped, having something like example 7.1 in mind.
[3] See also Fine (1985b, p. 34).

set I of individuals. Such objects may be generated in stages, according to their 'degree' of independence. At the first stage are the independent objects. Since there are no essential constraints on the existence of arbitrary objects, we should expect that to each set of individuals from I there will be an arbitrary object with that set as its range. At the second stage are the arbitrary objects that depend upon the objects generated at the first stage, but not on anything else. ... For each set I of individuals, we thereby obtain a 'system' A_I of arbitrary objects.

But this seems too restrictive.[4] Fine himself admits that it seems perfectly fine to start a conversation by saying (Fine 1983, p. 70):

Example 7.3.
Take two arbitrary real numbers a and b. Now consider their sum $a+b$ ...

This suggests that there are at least two *independent* arbitrary reals a and b. Fine himself considers a way of disabusing oneself of this view by making sense of the example without assuming more than one independent real number (Fine 1983, p. 70), but I will not go into his suggestion here.[5]

This identity condition for arbitrary objects is reiterated in (Fine 1985b, p. 34), although Fine goes on to add that 'for certain technical purposes, a smoother theory is obtained by the posing that ... many different [variable objects] may conform to the same identity criterion' (pp. 34–35). But it seems that the foregoing example shows that more than technical reasons may be adduced in favour of allowing more than one independent variable objects of a given kind: it just seems to be the natural and straightforward thing to do.

Fine (1998, p. 610) later proposes the following criterion of identity for *systems of variable objects*:

Let S and S' be any two systems of variables; and let us say that S and S' are similar under a one-one map $x \rightarrow x'$ from the variables of S onto the variables of S' if the values taken by the variables of S are the same as the values taken by the corresponding variables of S'. Thus, if S consists of the variables $x_1, x_2, \ldots$ and S' of $x'_1, x'_2, \ldots$, then the systems will be similar if, for any objects $a_1, a_2, \ldots$, $a_1, a_2, \ldots$ will be the respective values of $x_1, x_2, \ldots$ just in case they are the respective values of $x'_1, x'_2, \ldots$ The criterion of identity then states

---

[4] This claim is also (tentatively) made in Santambrogio (1988, p. 634).
[5] Scepticism about Fine's proposal is expressed by Macnamara in his review of Fine's book (Macnamara 1988, p. 306).

that the variables of similar systems are the same. The identity of a system of variables is given by the values that the variables assume.

This principle allows for systems with more than one independent variables. But this principle seems incorrect by Fine's own lights. We have seen in the previous section that Fine took example 7.1 and example 7.2 as describing to *different* systems of variables, because the dependence relation in them is different in nature. But the criterion of identity for systems of variable objects forces us to identify these two systems. We could turn this on its head and see this as evidence that example 7.1 and example 7.2 *cannot* describe different systems after all. However this would be to give foundational metaphysics precedence over naive metaphysics: Fine would surely object to this.

7.3.2 Systems of variable objects. There are at least two kinds of uses to which arbitrary objects have been put. Firstly, arbitrary objects play a role in reasoning patterns. Secondly, arbitrary objects perform representational functions.

Fine is occupied with both of these uses of arbitrary objects. His early work on arbitrary objects, especially Fine (1985a, 1985b), is first and foremost concerned with reasoning with arbitrary objects. Much of this work focuses on describing natural and perspicuous *proof systems* for reasoning with arbitrary objects. His more recent writings on arbitrary objects, especially Fine (1998, 2017a), pertain to representational uses of arbitrary objects. But it seems to me that Fine's earlier interest in reasoning with arbitrary objects has informed his *theory* of arbitrary objects in certain respects that perhaps it should not have.

It is reasonable to expect that in any given mathematical argument – say a number theoretic one – only finitely many arbitrary objects are introduced. Moreover, the argument will wear the way in which these arbitrary objects depend on each other on its sleeve. So for the purposes of modelling reasoning, it suffices to consider models with a finite number of arbitrary objects, arranged in a *system*. But Fine's work on mathematical structure (see Section 7.5) shows that for certain representational uses of arbitrary objects, systems containing an infinite supply of arbitrary objects need to be taken into account.

A distinctive feature of Fine's theory is that he *only* discusses arbitrary objects in the context of, and as belonging to, a *system* of arbitrary objects. As far as I can see, Fine leaves it open whether there is a *universal* system consisting of all arbitrary objects.

Fine's discussions of arbitrary objects always have a *local* feel: most of the systems that he considers contain less than a handful of arbitrary objects. In Fine's framework, one must be careful with drawing *global* conclusions based on discussions of local systems. For instance, it is easy to see that a system S of arbitrary objects may contain two disjoint subsystems S_1 and S_2 such that when Fine's criterion of identity for systems is applied to S_1 and S_2 individually, the two systems are *identified*.

7.3.3 Comprehension.

There is a *naive* comprehension principle that at first sight appears to be basic and to capture a fundamental aspect of the nature of arbitrary objects. Fine (1983, p. 59) calls it the *principle of generic attribution*: 'the principle that any arbitrary object has those properties common to the individuals in its range'. In other words, when taking a restriction to some value range for variable objects to be implicitly given, we have

Thesis 7.4 (Generic Attribution).
For any independent variable object, call it a, and for any property φ:

$$\varphi(a) \Leftrightarrow \forall i : \varphi(i),$$

where the quantifier $\forall i$ ranges over the specific φs in the range of a. Of course there is a version of this principle for φs with more than one argument place, and Fine considers it, but we will stick to the simple version here.

Thesis 7.4 is widely taken to be refuted by the following argument, which seems due to Berkeley (1710, Introduction, X). I here formulate it for a particular kind of objects (the natural numbers), but it is clear that the line of reasoning is general:

Let F be 'natural number', and let a be the (or 'a', if you think there are more than one) independent natural number. The arbitrary natural number a is a natural number. Every natural number is either even or odd. So a is either even, or odd. But by the Generic Attribution *Thesis 7.4, a is neither even, nor odd. Contradiction.*

Another argument for the same conclusion is usually ascribed to Leśniewski. A reconstruction of this argument is given in Bacon (1974), but I here follow the smoother exposition of Santambrogio (see Santambrogio 1987, p. 639). Suppose that, given a predicate $\varphi(x)$, we can form a *term*

$\vartheta x\varphi(x)$, which stands for some arbitrary φ.[6] Then we can reason as follows. Let $Q(x)$ be the property of being identical with $\vartheta x\varphi(x)$, i.e.

$$Q(x) \equiv (x = \vartheta x\varphi(x)).$$

The left-to-right direction of the principle of generic attribution says that:

$$Q(\vartheta x\varphi(x)) \rightarrow \forall x(x = x \rightarrow Q(x)).$$

By the way in which we defined the predicate $Q(x)$, we can spell this out as

$$\vartheta x\varphi(x) = \vartheta x\varphi(x) \rightarrow \forall x(x = x \rightarrow x = \vartheta x\varphi(x)).$$

But by the laws of identity, the latter entails $\forall x : x = \vartheta x\varphi(x)$, which is patently absurd.

Fine agrees that the principle of generic attribution, as stated by Thesis 7.4, is not correct. Fine therefore tries to solve the problem by finessing Thesis 7.4. He distinguishes between the *generic reading* of a condition and a *classical reading* of a condition (Fine 1983, pp. 63–65). The idea is that a typical condition of numbers – but this generalises to any other kind of objects – can be taken either as applying to all arbitrary numbers (this is the generic reading), or only to specific numbers (this is the classical reading). Fine (1998, p. 612) gives the following example:[7]

Let us suppose that the variable sign **x** *signifies the variable object x. Then, under a generic reading, the assertion*

x *is a variable object*

will be true only if each value of x is itself a variable object. But under the literal [i.e. classical] reading, the assertion will be true, regardless of what values are assumed by x.

Fine (1983, p. 64) then restricts the principle of generic attribution (Thesis 7.4) to *generically read* conditions. On the generic reading, the arbitrary number a in Berkeley's argument is indeed either even, or odd, so there is no problem.

---

[6] Santambrogio supposes that $\vartheta x\varphi(x)$ stands for *the* arbitrary φ, but this is not essential for the argument.

[7] In the journal article, this passage is somewhat difficult to decipher because no distinction in font is made between the expression for the variable sign (**x**) and the expression for the variable object (x).

For many predicates, the generic reading is not the natural one. Take for instance the predicate

$$x \text{ is a specific object.}$$

If we leave higher-order variable objects aside, then the condition expressed by the predicate on the generic reading is *trivial*: it holds of every arbitrary object. But in variable object theory, this condition should not come out as trivial. Rather, it marks a fundamental metaphysical distinction between two kinds of objects. So the classical reading of this predicate seems the right one, and this reading is not covered by Thesis 7.4.

Some commentators find Fine's distinction between generic and classical (readings of) conditions difficult to understand or problematic (Tennant 1983; Breckenridge and Magidor 2012, pp. 389–390). Fine himself recognises that this distinction is a somewhat subtle and difficult part of his account. At any rate, it seems somewhat awkward to postulate, as Fine does, a fundamental *ambiguity of meaning* of expressions. It would be desirable to have a language with *exactly one* intended interpretation, in which every generic and every classical condition can be expressed in a natural manner.

7.4 Cantorian Abstraction

Fine aims to formulate a theory of *abstraction* as the process of freeing an object of its peculiar features, which yields a conception of number or order type as the product of such a process (Fine 1998, p. 600). In particular, his aim is to give a natural account of Cantor's and of Dedekind's abstraction processes, where Cantor abstracts finite natural numbers from sets of objects, and Dedekind abstracts finite order types from ordered systems of objects. Fine's account relies on his theory of arbitrary objects. The relation of *dependence* between arbitrary objects plays a crucial role (Fine 1998, p. 613).

We first consider Fine's theory of Cantorian abstraction of finite cardinal numbers. The number two, for instance, is abstracted from sets (unordered pairs) of specific numbers. Let us see what the numbers 1 and 2 are on his account: then it will be clear how the story goes for the other finite cardinals.

Consider an independent variable number u_1^1. Then, simply, the set $\{u_1^1\}$ is the number 1 (Fine 1998, p. 612).

Take two arbitrary objects u_1^2 and u_2^2 that only depend on each other, and that are only subject to the condition that $u_1^2 \neq u_2^2$. Note that the variable objects u_1^2 and u_2^2 both differ from the variable object u_1^1. Then

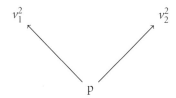

Figure 7.1 Dependence: order type 2.

the set $\{u_1^2, u_2^2\}$ is the natural number 2 (Fine 1998, pp. 612–613). Observe at this point that by Fine's criterion of identity of objects in systems, the unordered pair is *unique*. Observe also that the number 2 is then not a variable object, but a *set*. So if being instantiated means taking a value, then the number 2, on the proposed account, cannot be said to be instantiated by the doubleton of Mount Fuji and Bristol Suspension Bridge, say.

Next, let us look at Fine's theory of Dedekind abstraction of finite order types. The nature of the order type 2 is slightly more complicated than the nature of the natural number 2.[8] Consider an independent variable object p that has as values all systems of the form $\{\langle a, b\rangle\}$, where the ordered pair $\langle a, b\rangle$ is taken to be defined in a standard set theoretic way (by means of Kuratowski's definition, for instance). Fine calls the object p the *prototype* of the type that he wants to construct (Fine 1998, p. 618). Let v_1^2 be the dependent variable object that always takes the value of being the *first* element in any given value of p; let v_2^2 be the dependent variable object that always takes the value of being the *second* element in any given value of p. The dependence structure between p, v_1^2, v_2^2 is depicted in Figure 7.1. Then the order type 2 is the set $\{\langle v_1^2, v_2^2\rangle\}$ (pp. 617–618). Note that $\{\langle v_1^2, v_2^2\rangle\}$ is not *identical* with p, for p is not itself of the form $\{\langle a, b\rangle\}$. By Fine's criterion of identity for variable objects in systems, p is unique, and therefore also v_1^2 and v_2^2 are unique.

Fine (1998, p. 623) distinguishes between *representational* and *non-representational* accounts of types (kinds) of objects:

Following Hallett ..., we shall say that an account of the types of some kind is representational if each type of the given type is of that very type. The accounts of cardinal number given by von Neumann and Zermelo are representational in this sense: each cardinal number is of that number ... On the other hand, the accounts of cardinal and ordinal number of Russell and Frege are not

---

[8] I assume that I won't cause confusion by re-using the symbol '2' to refer to the order type.

representational: each cardinal or ordinal number will not (as a rule) be of that number.

Fine (1998, p. 624) gives a cautious philosophical defence of representationality as a feature of accounts of types:

> There may also be a certain philosophical advantage in adopting a representational theory; for certain types do seem to be self-applicable. We do seem to conceive of the order type of a ordering, for example, as an ordering that is of that very type, though whether a cardinal number should be regarded in a similar same way as a representative item of that number is not so clear.

The natural number 2 and the order type 2, as defined above, are clearly representational types.

7.5 The Natural Number Structure

Fine (1998, Section VI) applies his approach to aspects of the mathematical structuralism debate.[9]

He proposes the following way of defining the natural number structure (p. 630):

> It is here that our account of Cantorian abstraction may be of some help. For we may define a representational Cantorian type of a class of isomorphic models along the lines of our definition of order type. In the case of arithmetic, for example, the number 0 will be the first element in an arbitrary ω-progression, the number 1 the second element, and so on. Or, to be more exact, the prototype in this case will be an independent variable N whose values are all of the ω-progressions, the numbers $0, 1, \ldots$ will be variables $n_0, n_1, \ldots$, dependent upon N, whose values for any particular value $\underline{N}$ of N are the first, second, $\ldots$ members of $\underline{N}$, and n will be the successor of m if, for any value $\underline{N}$ of N, the value of n is the successor of the value of m in $\underline{N}$. Moreover, given such a definition, it may then be shown that the type is unique and itself a member of the class of models.

Thus we obtain the dependence diagram of Figure 7.2.

And then the natural number structure is taken to be the ω-sequence

$$\langle n_0, n_1, n_2, \ldots \rangle.$$

---

[9] Fine does not endorse the resulting flavour of structuralism (Fine 1998, p. 630).

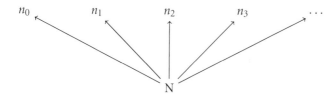

Figure 7.2 Dependence: the natural number structure.

Clearly this approach can be applied to other individual mathematical structures. The abstractive movement involved in this construction is of a somewhat different kind than the Cantorian abstraction: Fine calls it *Dedekindean abstraction* (Fine 1998, p. 630).

Note that by Fine's criterion of identity for variable objects in systems (Section 7.3), there is a *unique* independent variable object N on which the construction is based.

One should not take Fine's account to be too close to Shapiro's non-eliminative version of structuralism. On Fine's proposal, the natural number structure is taken to be the *set* $\langle n_0, n_1, n_2, \ldots \rangle$ rather than the prototype N, which is what I called the *generic ω-sequence*.

The *set* $\langle n_0, n_1, n_2, \ldots \rangle$ is a representational type. But whether it is a structure in Shapiro's sense of the word, is not clear. It would be natural to take having a value to be the way of making the instantiation relation precise. But $\langle n_0, n_1, n_2, \ldots \rangle$ is not a variable object, so it is not the kind of entity that takes values. So, in particular, this entity is not the same as the generic systems that I took mathematical structures to be in Chapter 6.

Fine argues that this Cantorian theory of abstraction of finite and infinite types has advantages compared to existing theories of mathematical objects and structures.

Frege objected against the *phychologism* inherent in Cantor and Dedekind's accounts of abstraction (Frege 1984, Sections 28–44). Fine accepts Frege's critique, and argues that the above abstractionist account of types contains no trace of psychologism (Fine 1998, pp. 603, 616).

Fine's account of the natural number structure is *representational*. In this respect, it is in agreement with Shapiro's non-eliminative structuralism about the natural number structure. The reason is that on Fine's view, there is a salient sense in which the natural number structure instantiates itself. In this respect, Fine's view of mathematical structures is *unlike* the account of the natural number structure that was proposed in Chapter 6.

We have seen in Section 5.3 that set-theoretic reductionist accounts of the natural numbers, while representational, have been accused of containing

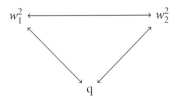

Figure 7.3 Dependence alternative: order type 2.

a degree of *arbitrariness*. And we have seen in Section 5.5.2 how Hellman's permutation objection shows that Shapiro's version of non-eliminative structuralism is also vulnerable to this objection.

Fine argues that his Cantorian account of the individual natural numbers combines the best of both worlds. It is representational, and therefore superior to Frege and Russell's account of natural number. But it is also non-arbitrary (Fine 1998, pp. 627–628), and therefore superior to von Neumann and Zermelo's accounts of the natural numbers.

Fine's argument for the non-arbitrariness of his abstractionist view rests upon the thought that his account attributes to each type (natural number, for instance) exactly those properties that instantiate it, and nothing more (Fine 1998, p. 625). On the cardinality conception, this is captured by equinumerosity. We have seen that Fine's criterion of identity for variable objects in systems guarantees that the systems in question are unique.

Fine acknowledges the following objection against his claim of non-arbitrariness. We have seen how Fine's abstractionist account of the order type 2 involves *three* variable objects: one independent variable object (p), and two variable objects (v_1^2 and v_2^2) that depend only on p. A different system consists of three variable objects (q), and two variable objects (w_1^2 and w_2^2) that all co-depend on each other (see Figure 7.3).

Why is the order type 2 the ordered pair based on the first system rather than on the second system? Fine (1998, p. 629) responds as follows:

> There is, however, something counterintuitive about taking the variables to be codependent (even though, for reasons of technical simplicity, we may treat them as if they were). For we think of [v_1^2 and v_2^2] as the constituents that occupy the first and second positions in a given ordered pair. Thus we take what they are to be dependent upon what the given ordered pair . . . happens to be, and not vice versa. Moreover, this intuitive view is in line with the general policy of only assuming those dependencies required for the purposes at hand.

Fine does not elaborate on whether and how his account of the natural number structure avoids the arbitrariness objection. But one imagines that his defence goes along similar lines.[10]

Nonetheless, the critic will maintain that for all mathematical purposes, the first system (Figure 7.1) serves as well as the second (Figure 7.3). So one may wonder whether Fine's account of the order type 2, for instance, does not after all contain excess structure. In sum, I am not convinced that the arbitrariness objection has been completely disappeared.

A similar point can be made about Fine's abstractionist theory of the natural number structure. But in any case, we are left with one further question concerning the natural numbers. The Cantorian natural number 2 is not the same entity as the number two in the structure of the natural numbers as proposed by Fine. How do they relate to each other? Which of these two entities, if any, is the *real* number 2? The answer to this question presumably depends on whether structuralism about arithmetic is basically the right stance to take: this is not a question that is addressed in Fine (1998).

7.6 Predicativism?

Fine developed a *postulational* theory of mathematical objects (Fine 2005b). Without going into details, the idea is that mathematical objects are brought into the reach of our quantifiers by *imperative postulates*, which can be seen as instructions to construct objects with specified properties. The imperative character of these postulates distinguishes them from the *indicative* postulates (i.e. axioms) involved in abductive strategies that are used in empirical science, for instance. If we take a postulational line, then epistemological questions about arbitrary objects become in a sense trivial. Our knowledge of them is postulational: they have the properties that they are posited to have.

On the other hand, Fine (2005a) has formulated a theory of *classes* according to which classes are given from the outset, and not stipulated into existence. Fine himself recognises the fundamental tension between the two approaches (p. 572 n23). Indeed, he shows that accepting quantification over absolutely all mathematical objects is incompatible with his

---

[10] Fine states that Section VI of Fine (1998) was substantially condensed at the behest of the editors of the journal (p. 629 n22). It would not surprise me if this point was addressed in the original, unabridged, version of the article.

postulational theory of mathematical objects (Fine 2005b, Section 3). In sum, in his philosophy of mathematics, Fine vaccillates between a predicative and an impredicative approach.

In the development and defence of his postulational theory, Fine makes no reference to arbitrary objects. But his theory of arbitrary objects has a distinct postulational flavour: it is tempting to interpret him as seeing arbitrary objects as being brought into the reach of our quantifiers by an act of postulation.

This is why the logical analysis of ordinary language plays a central role in Fine's account. We have seen in the present chapter that Fine is mostly interested in situations where a finite number of arbitrary objects are gradually introduced in a context (in a proof, for instance). Moreover, the clause

$$\text{'LET } a \text{ be...'}$$

of Fine (1998), in which arbitrary objects are introduced into a conversation, certainly appears to be a form of postulation. But postulates intend to *define* entities. So I tentatively conclude that Fine implicitly advocates some kind of predicativism about arbitrary objects.

In this matter, Fine's view contrasts sharply with mine. My metaphysical account of arbitrary objects is resolutely impredicative. I take arbitrary objects to be encountered by us in our metaphysical experience of the world rather than somehow to be postulated into existence. I take all arbitrary objects to be given all at once, and take absolutely general quantification over them to be perfectly legitimate.

Our metaphysical experience of the nature of the world is bound to leave traces in our language. So if you want to conduct an enquiry into the nature of arbitrary objects, natural language is a good place to start. Nonetheless, in my view, natural language considerations are ultimately of limited relevance in the theory of arbitrary objects: the nature of arbitrary objects can not be fully read off from an analysis of the 'grammar' of our language (to use a Russellian phrase). And certainly we cannot conclude that arbitrary objects are always given by predicates.

Because of the impredicativity of arbitrary objects, epistemological questions about them cannot be deflated in the way that Fine seeks to deflate questions of existence of mathematical objects. In my view, epistemological questions about arbitrary objects are anything but trivial (see Sections 4.4 and 4.5): they concern the way in which the nature of the world is given to us.

7.7 Universals versus Variable Numbers

So far in this chapter, I have contrasted Fine's theory of arbitrary objects with mine. It is tempting to ask: which of these two arbitrary object theories, if any, is basically correct? However, it should not be taken as a given that there is only one concept of arbitrary object; perhaps Fine and I are zooming in on related but somewhat different concepts. Let me explain.

Platonistic universals were posited at least in part to explain our ability to acquire *general knowledge*. According to this view, we obtain general knowledge by being in cognitive contact with ideas or Forms in which specific objects of a given kind somehow *participate*. The Forms then do not have all the properties that are shared by the sensible objects that instantiate them. For instance, the idea of extension is abstract, and arguably not itself extended.[11]

One basic feature of platonistic theories of universals is that for each multitude of objects, there is at most *one* universal that they and only they participate in. The nineteenth century conception of variable is fundamentally different in this respect. Indeed, the relation 'being a value of' differs *logically* from that of participation. There are *multiple* variable entities having the same value range; their values are correlated with each other in multiple ways. Therefore it is doubtful that one arbitrary object theory can yield both a satisfactory account of platonistic universals *and* a satisfactory account of the nineteenth century conception of variable numbers.

It seems to me that Fine's theory and also Santambrogio's theory provide frameworks in which intriguing accounts of platonistic universals can be developed. Indeed, this is what they intend to do. Santambrogio developed in some detail an affirmative answer to the question: can we 'objectify' a universal such as a kind or a species into its generic representative or its typical member?[12] In Fine (2017a), Fine formulates a theory of logical form of formulas of sentential logic. But this is only by means of illustration; it is merely a pilot study. He takes it as a more substantial aim of his arbitrary object theory to develop a theory of the more concrete *types* of ordinary and scientific discourse.

I am sceptical that the theories by Fine and Santambrogio form a suitable setting for a rational reconstruction of the nineteenth century conception of variable numbers. Indeed, there is a tension between on the one hand

---

[11] This relates to the philosophical debate concerning Plato's view of sentences such as 'Beauty is beautiful': see Mates (1979), for instance.

[12] Santambrogio (1987, p. 638).

arbitrary object theory as I develop it, and on the other hand a theory of universals in Fine's and Santambrogio's sense.

In an effort to explain this tension, let us go back to the ordinary language sentence

The cow is a mammal.

that was discussed in Section 3.1. There are many arbitrary cows. Indeed, there are many diagonal cows. But no arbitrary cow is the (unique) generic cow – or so I maintain. If we do maintain that some arbitrary cow is the generic cow, then we open ourselves to a Benacerafian identification problem: why is *this* arbitrary cow the generic one rather than *that* one?

Fine might in reply say that there is only one *independent* arbitrary cow: she is the winner! But this seems to shift the problem rather than to solve it. Why is *this* arbitrary cow the one and only independent arbitrary cow rather than *that* one?

Perhaps at this point Fine's postulationalism provides a way out: perhaps the generic cow was propelled into existence, as it were, by an act of postulation. But I find it difficult to reconcile postulationalism about arbitrary objects with the metaphysical appearance of arbitrary objects as given 'all at once' and prior to any acts of postulation.

However this may be, I hope that the arbitrary object theory that I try to articulate in this book *is* a natural framework for reasoning about variable numbers in the nineteenth century sense of the word; I do not claim that it is a suitable setting for a theory of universals.

Generic Systems and Mathematical Structuralism

> Ideas from the metaphysical tradition can be misleading when imported into discussions of mathematical structuralism and perhaps into discussions of mathematical objects generally.
>
> *(Parsons [2004, p. 74])*

Are mathematical structures *sui generis* entities, or is talk about mathematical structures reducible to talk that does not mention mathematical structures? This incarnation of the realism debate takes pride of place in the contemporary philosophical debate about mathematical structuralism.

As a naive metaphysician, I find it unclear what is at stake in this debate, and I want to move away from this discussion in the philosophy of mathematics. It seems to me more fruitful to reflect on the nature of mathematical structures.

Eliminative and non-eliminative forms of mathematical structuralism nonetheless contain insights that every decent philosophical account of mathematical structures has to accommodate. So the challenge is to incorporate the insights in the nature of structure that have been acquired by eliminative and non-eliminative mathematical structuralists and to do better where existing accounts have fallen short. It is in this spirit that I connect my account of mathematical structure with elements of leading versions of mathematical structuralism.

8.1 Generic Structuralism?

The background is now in place against which the philosophical account of the nature of mathematical structures that I want to advocate can be articulated. I am sometimes tempted to dub this account *generic structuralism*. But on reflection, I think that doing so would be a mistake. The account that I propose is not comfortably situated as an '-ism' among the existing forms of mathematical structuralism.

My philosophical account does not aim to do all the things that existing forms of mathematical structuralism try to do. The naive metaphysician is not interested in trying to eliminate reference to and quantification over mathematical structures, or to reduce it to something else. Nor is she interested in arguing that mathematical structures 'really exist'. The theory that is advocated in this book does not claim that mathematics is only, or even primarily, about mathematical structures. It does not claim, for instance, that the subject matter of set theory is one or many mathematical structure(s); it does not claim that computability theory is about one or many mathematical structure(s). But it *does* claim that mathematics is concerned, among other things, with mathematical structures.

I accept the perhaps somewhat vague distinction between algebraic and categorical mathematical disciplines. I agree with Parsons that categorical mathematical disciplines theorise about *one* mathematical structure and about the *mathematical* objects that belong to this mathematical structure.

As far as algebraic mathematical disciplines are concerned, we have seen in Sections 6.7 and 6.8 that there are two options.

One option is to say that countable graph theory, for instance, describes the arbitrary countable graph. On this view, the distinction between algebraic and categorical disciplines is not as fundamental as it seems. They both describe a unique structure; their difference lies in the fact that an algebraic discipline describes a structure that does not clearly contains particular mathematical objects.

But we have seen that this is not the natural way to go. Isaacson was right to insist that even for algebraic disciplines, it is the individual structures that are fundamental (Section 5.4.3). The generic countable graph, for instance, is too 'attenuated' convincingly to serve as the structure that is investigated by countable graph theory. So, according to my account of mathematical structure – as according to Shapiro's version of non-eliminative structuralism – algebraic mathematical theories purport to describe a *family* of mathematical structures that are united in making a fixed list of axioms true. All this does not mean, of course, that structures like the arbitrary countable graph might not be of metaphysical or indeed of mathematical interest.

I have argued that a mathematical structure is a *generic system*. This system is an arbitrary entity that can be in the state of being this or that particular system.

On my conception, generic systems are not self-instantiating. In the previous chapter we saw how this distinguishes the account of mathematical structure that is developed in this book from the account that was developed by Fine. Fine's view is closer to Dedekind's conception of structure.[1]

---

[1] Thanks to Oliver Tatton-Brown for pointing this out.

Consider, for instance, the following famous passage in Dedekind (1888, paragraph 73):

If, in considering a simply infinite system N [i.e. an ω-sequence], ordered by a mapping φ, one abstracts from the specific nature of the elements, maintains only their distinguishability, and takes note only of the relations in which they are placed by the ordering mapping φ, then these elements are called natural numbers *or ordinal numbers or simply* numbers, *and the initial element 1 is called the initial number of the* number series N.

It seems clear that Dedekind's number series N is then itself an ω-sequence.

I have argued that mathematical structures, as generic systems, contain mathematical objects. These mathematical objects are themselves arbitrary objects. We have seen in Section 6.8 that in the case of structures admitting non-trivial automorphisms, the mathematical objects that they contain are *second-order* arbitrary objects: objects that can be in states that themselves are arbitrary entities.

Consider first rigid structures. As a representative example, let us look again at arithmetic. In the setting of arbitrary objects and systems, sense can be made of the ordinary statement that arithmetic is about the natural numbers. In fact, there are two different ways in which this can be made more precise. You might say that arithmetic is about all *specific* natural numbers; alternatively, you might say that arithmetic is also about *arbitrary* natural numbers.

It seems incorrect to say that arithmetic is about the specific *and the arbitrary* natural numbers. After all, mathematicians only prove theorems about the specific natural numbers. But this is not a problem for the account that I am proposing. To say that arithmetic is about the generic system **N** (of Section 6.4) does not commit you to holding that it is (among other things) about arbitrary natural numbers.

What matters from a metaphysical point of view is that the arbitrary natural numbers are 'there' in the generic system **N**, waiting to be discovered and investigated. Even the specific natural numbers are, on the view that I am proposing, arbitrary objects. Once arbitrary objects are taken metaphysically seriously, arbitrary natural numbers must be taken seriously also; we *find* arbitrary numbers in the natural number structure.

At any rate, I do not believe that we can lean heavily on analysis of ordinary discourse for support of the theory of arbitrary natural numbers. As an illustration, consider a famous conjecture:

Thesis 8.1 (Goldbach). Take any even natural number larger than 2; it is the sum of two prime numbers.

Its generic structuralist readings are as follows:

1 Take any specific even natural number n; then n is the sum of two (specific) primes.
2 Take any arbitrary even natural number n; then n is the sum of two (arbitrary) primes.

In Fine's terminology (see Section 7.3.3), the first is the 'generic' reading of Goldbach's conjecture; the second is its 'classical' reading.

The first reading is clearly the correct one. But this is of little significance to our present endeavour. At some point in our metaphysical investigation we must leave informal mathematical and non-mathematical discourse behind. As a naive metaphysician, I am happy to concede that a Frege–Russell analysis of the logical form of Thesis 8.1 in terms of quantification over *specific* objects is the only right one. This takes nothing away from the metaphysical importance of the theory of arbitrary systems and arbitrary numbers.

8.2 Themes from Eliminative and Non-Eliminative Structuralism

Let us now turn to the comparison between the account of structure that was articulated in Chapter 6 and forms of mathematical structuralism. I argue that the view of mathematical structures as generic systems incorporates the insights concerning the nature of mathematical structures that were reached by eliminative and non-eliminative forms of structuralism. Moreover, I contend that my account of generic systems is not vulnerable to known objections against existing forms of mathematical structuralism. In particular, the objections of Section 5.5.2 against Shapiro's non-eliminative form of structuralism do not apply to the generic conception of mathematical structures.

8.2.1 Instantiation and modality. Non-eliminative structuralists stress that *ante rem* structures can be *instantiated* by systems of objects (Shapiro 1997). Eliminative structuralists emphasise that systems of objects *play the role* of mathematical structures (Benacerraf 1965). Moreover, some eliminative structuralists insist that all that matters for mathematical structuralism is which systems of objects are *possible* (Putnam 1967; Hellman 1993).

Generic systems are governed by an instantiation relation: the relation of *being in a state*. This is what makes my view akin to a *structuralist* view in the philosophy of mathematics despite the fact that I do not have much to say about the realism question. Moreover, the notion of generic system recognises an implicit *modal* component in the notion of mathematical structure: mathematical structures are entities that *can be* in states.

If the generic structuralist – if we may call her that – is right, then for a system S to instantiate a structure **S** means no more than that **S** can be in the state of being S (but is not numerically identical with S).

There is little more to say about what it means for a system S to instantiate a structure **S** than S can be in the state of being **S**. This means that instantiation is a very ethereal relation, and this may in turn be be a partial explanation of the attractiveness of the eliminativist view. The reason is that it is tempting to continue this line of reasoning as follows. If **S** is not $S, T, U, \ldots$, then it is nothing in our world at all, for there are only objects and entities (or pluralities) such as $S, T, U, \ldots$ But this last statement is the result of an anti-realist and foundational inference that goes beyond how mathematical structures present themselves to us. For this reason, this statement goes beyond what the naive metaphysician is prepared to assert (or deny).

Shapiro (1997, Chapter 7) has argued that there is nothing to be gained by invoking a modal notion in the interpretation of mathematical discourse. He even seems cautiously to express sympathy for the thesis that modality can be *reduced* to set theory (pp. 2018–2019):

I show that there are straightforward, often trivial, translations from the set-theoretic language of the realist to the proposed modal language and vice versa. *The translations preserve warranted belief, at least, and probably truth … The contention is that, because of these translations, neither system has a major epistemological advantage over the other. Any insight that modalists claim for their system can be immediately appropriated by realists and* vice versa. *Moreover, the epistemological* problems *with realism get 'translated' as well …*

Where, then, is the burden of proof between the schools? Other things being equal, it would be nice to take the languages of mathematics, set theory in particular, literally.

Nominalist philosophers of mathematics disagree. Hellman (1993), as a *nominalist* modal eliminative structuralist, makes use of a concept of con-crete possibility in his explication of the possibility statements that guarantee the non-vacuousness of ordinary mathematical statements. We have noted in Section 5.4 that, whereas it seems plausible that there might have been a concrete system that forms a model of arithmetic, it is not clear that there could have been a *concrete* system that forms a model for set theory, for instance.

Field (1984), in his fictionalist account of mathematics, makes use of *logical* possibility to explicate the sense in which it is a requirement of mathematical 'stories' to be consistent or coherent. But here Shapiro might argue that logical possibility is underwritten by mathematical existence (existence of a model). Parsons agrees with Shapiro. Indeed, it is on such grounds that Parsons suspects that in the end logical possibility reduces to *mathematical possibility*: what for all mathematics can tell us can be.

Parsons takes mathematical possibility to be a *sui generis* modality. He argues that it cannot be identified with *metaphysical possibility* (Parsons 2008, Section 15). Metaphysical possibility and impossibility derive from the *nature* of the entities, properties, and relations involved. The modal presuppositions that are supposed to guarantee the non-vacuity of our mathematical theories are simply not of this kind.

The modality that is connected to arbitrary objects is different. Possibility to be in a certain specific state (afthairetic possibility) *does* derive from the nature of the object in question. In this respect, afthairetic possibility resembles metaphysical possibility. So afthairetic possibility cannot be identified with mathematical possibility as Parsons conceives it. Nonetheless, I have also argued earlier (in Section 3.6) that afthairetic possibility is not *the way our world might have been as far as metaphysics is concerned*. In sum, if I am right, then none of the main structuralist views (Field, Hellman, Parsons) gives an accurate account of the notion of possibility that is involved in the notion of mathematical structure.

8.2.2 Mathematical objects.

According to the eliminative structuralist, there are no mathematical structures. Talk about mathematical structures can be *reduced* to talk about systems. There are then also no mathematical objects (Section 5.4.3): if it is not true in a naive sense that there is a natural number structure, then it cannot be true in a naive sense that the number three belongs to the natural number structure. The closest we can come to saying that the number three belongs to the natural number structure is to say that (necessarily), for every ω-sequence S, there is an object in S that is the third element in the ordering of that system. In a word, apparent direct reference to specific numbers is regarded as implicit universal quantification over entities of some background domain.

In Section 5.4.3 I briefly discussed how eliminative structuralists explain the seeming behaviour as names of expressions such as '$\mathbb{N}$', '$\mathbb{Q}$', '6' in terms of the concept of *dedicated variable* (Pettigrew 2008). The distinguishing feature of a dedicated free variable is that when it is assigned a denotation, it cannot on another occasion be re-used to refer to another entity. One can in this context distinguish between *independent* dedicated variables, such as 'N', and *dependent* dedicated variables, such as '6'. The point, however, is that *implicitly*, a dedicated variable such as 'N' is implicitly universally quantified over.

Non-eliminative structuralists object to this account on broadly naturalist grounds. They argue that the eliminative paraphrases give a rather convoluted explanation of the data, i.e. of what mathematicians say. We have seen how they go on to add that their own view allows us to interpret what mathematicians say *literally*[2] and therefore explains it better.

Some work has been done to compare Fine's (1985a) logical treatment of variables with logical treatments of variables in the spirit of Pettigrew's designated variable account: see King (1991). I will not go into details here: I suspect that as far as natural language semantics goes, what one account can do, the other can do, too – even though it may be the case that one of them will the more elegant one.

I agree with Shapiro that expressions such as 'N' and '6' function as names, but I add: names for arbitrary entities. However, as a naive metaphysician, I am unhappy with seeing the problem in terms of *explanation*. I take formal and informal discourse not as data to be explained, but as an expression of how the nature of mathematical structures presents itself to us: mathematical structures present themselves as consisting of *objects*. More is at stake here than having an ontology that lends itself to giving an elegant representation of informal mathematical discourse. Taking a naive metaphysical attitude towards mathematical structures may lead us to discover a rich and fruitful metaphysical theory that is intrinsically valuable.

We have seen in Section 5.5.2 how Shapiro's non-eliminative structuralism is open to a Benacerraf-style objection. If the natural number structure can itself be seen as a special ω-sequence S^* consisting of objects ('positions-as-objects') that stand in relationships to each other, then a non-trivial permutation of these objects results in a different ω-sequence that has as much right to claim to be the natural number structure than S^* has. But the full generic natural number structure $\mathbf{N}$ as described in Section 7.5 is not an ω-sequence. Mathematical structures conceived of as generic systems *do not* instantiate themselves. Thus there is no permutation problem for the view that I propose.

---

[2] At least to a large extent: see page Section 5.4.

Undoubtedly there exist many set-theoretical models that are different from the ones that I have produced in Chapter 6 and that do an equally good job at representing the salient features of generic structures. But it is important (see Section 2.7) not to confuse set-theoretical models of generic systems with the generic systems themselves. No class or set or plurality of sets at all is proposed in my account as *being* the natural number structure, for instance.

8.2.3 Incompleteness.
In his discussion of the incompleteness of mathematical objects, Shapiro (1997, p. 100) writes:

mathematical *structures are* freestanding. *Every office [i.e. position] is characterized completely in terms of how its occupant relates to the occupants of the other offices of the structure.*

This is an endorsement of an abstraction principle that is an analogue of the abstraction principle for structures that was discussed in Section 5.1. The *abstraction principle for positions* says that if S and S' are systems, and $p(x, S)$ $(p(x', S'))$ stands for the *position* occupied by x (x') when S (S') is seen as an exemplification of an *ante rem* structure (Linnebo and Pettigrew 2014, p. 274):

Thesis 8.2 (Abstraction for positions).

$p(x, S) = p(x', S') \Leftrightarrow [\exists \text{ isomorphism } f : S \to S' \text{ such that } f(x) = y].$

Since structures can in Shapiro's view themselves be regarded as systems, this entails that according to Shapiro's account we have for every *structure* S^*:

Consequence 8.3.

$p(x, S^*) = p(y, S^*) \Leftrightarrow [\exists \text{ automorphism } f : S^* \to S^* \text{ such that } f(x) = y].$

Mathematical structures that admit non-trivial automorphisms are the rule rather than the exception in mathematics. Recall, for instance, Burgess' example from Section 5.5.2 of the complex field $\mathbb{C}$, where i and $-i$ are related by an automorphism of $\mathbb{C}$. But since the position of a position(-as-object) *is* that very position, this entails that i and $-i$ are the same element of $\mathbb{C}$, which is patently absurd.

But even without the assumption that $\mathbb{C}$ instantiates itself, there is trouble. Consider two systems C_1 and C_2 that instantiate $\mathbb{C}$. Then there are two elements of C_1, call them i_1 and $-i_1$, that 'play the role' of i and $-i$ (in no particular order!). Similarly, there are then two elements of C_2, call them i_2 and $-i_2$, that 'play the role' of i and $-i$. The existence of isomorphisms $f, g : C_1 \rightarrow C_2$ such that $f(i_1) = i_2$ and $g(-i_1) = i_2$ entails by the laws of identity that $p(i_1, C_1) = p(-i_1, C_1)$. By similar reasoning we extend this to the conclusion that

$$p(i_1, C_1) = p(-i_1, C_1) = p(i_2, C_2) = p(-i_2, C_2).$$

Extrapolating from this, we see that *every* element i_C playing the role of one of i and $-i$ in some system C instantiating $\mathbb{C}$ determines the same position in $\mathbb{C}$: i or $-i$, but it is impossible to say which. Surely this is an unwelcome consequence.

Suspicion immediately falls on the abstraction principle for positions in a structure. But which abstraction principle should take its place?

The abstraction principle for positions that was implicit in Shapiro's quotation in the beginning of this subsection is a *criterion of identity* or *criterion of individuation* for positions in a structure. Quine (1948) insisted that for *every* kind of entities in our ontology we should have a criterion of identity: *no entity without identity!* The idea is roughly that if you do not have a criterion of identity for a putative kind of entities, then you do not really know what you are talking about when you are quantifying over entities of this kind, and hence you should not admit them in your ontology. Some philosophers accept Quine's methodological demand. Others reject it. Strawson (1976), for instance, argued that there is nothing wrong with admitting clouds in our ontology, even though any criterion of identity for clouds would contain a high degree of arbitrariness.

The above argument about the abstraction principle for positions in a structure has sparked a discussion in the literature (see MacBride 2005). A number of authors in the debate accept the Quinean requirement of a criterion of identity for positions in a structure. Some of them despair of the possibility of finding an adequate principle of individuation for positions and see this as an argument against non-eliminative structuralism (Keränen 2001); others attempt to find an adequate replacement for the principle formulated above (Linnebo and Pettigrew 2014; Schiemer and Wigglesworth forthcoming).

A number of authors join Strawson in rejecting the Quinean requirement for a principle of individuation for positions in a structure. Ketland (2006, p. 305), for instance, writes:

There still remains a more general philosophical problem, concerning the analysis of the notion of identity for the positions of an abstract structure. Must the identity relation on positions be defined in terms of the other distinguished relations? Or might the identity relation for positions be taken as primitive? For my part, I see no compelling reason why the identity relation, in general, should not be thought of as primitive.

Indeed, Shapiro himself later explicitly recanted not only any remarks in the spirit of the quotation at the beginning of this subsection – 'oops' was the technical term that he used in this context (Shapiro 2008, p. 286) – but also the requirement of an informative criterion of individuation for positions in a structure (Shapiro 2008, Section 2).

I agree with Strawson that it is not always the case that we do not know what we are talking about when we are talking about a kind K of entities unless we possess a criterion of identity for K. So I side with Ketland in not accepting Quine's *dictum* in general.

Preoccupation with a criterion of identity for a kind of entities has at least since Frege mostly been part and parcel of any ontological reduction attempt. Suppose, for instance, that we have arrived at a criterion of identity for a kind K of entities of the following form:

$$\forall x, y \in A : f(x) = f(y) \Leftrightarrow \Phi(x, y),$$

where Φ does not contain the function symbol f. Then there is bound to be someone who proposes that the Ks *really* are something like equivalence classes of As. Thus the search for identity criteria plays a role in *foundational metaphysics.*[3] But for a naive metaphysician who has little interest in foundational metaphysics, such considerations do not suffice to put the search for a criterion of identity for positions in structures high on the research agenda.

If, as might well be the case, it is impossible to find a plausible and informative criterion of identity for positions in mathematical structures, then we may expect a high degree of *indiscernibility* of positions in certain mathematical structures (Leitgeb and Ladyman 2008). Explicating and exploring the relevant notion of indiscernibility is then of interest in its own right. We will return to this in the next chapter.

Nevertheless, I do not wish conclude from the foregoing discussion that the notion of non-trivial automorphism is of limited interest. Indeed, we already witnessed the significance of this notion in Section 6.8, when we observed that the presence or absence of non-trivial automorphisms marks a difference in degree of abstraction of individual mathematical structures.

---

[3] This is in line with what Fine writes in Fine (2017b, Section 3).

At this point, we may ask whether the proposed view of mathematical structures gets the incompleteness of mathematical objects right. I believe it might. We have seen that on the proposed view, it does not make sense to identify the natural number 1 with any real number, for instance. We have seen that, in general, the objects that belong to generic systems are highly indeterminate: they are arbitrary objects. Nonetheless, MacBride argues, mathematical objects such as numbers must be taken to have the following 'non-structural' properties (MacBride 2005, Section 3):

1 Numbers are abstract.
2 Numbers have contingent properties.
3 Numbers can be used to count the number of elements in a collection.

We have seen that on the proposed account, numbers are indeed abstract entities. Also, it is not at all incompatible with the proposed account that the number 11 is your favourite number. And, to conclude, the natural numbers can be used to count the number of apples in a basket. On this latter point I have nothing original to propose. Counting the apples in a basket is simply a member of matching numerals with apples in an apple-exhausting way (not counting any apple twice): the last numeral 'gives' the number of apples.

8.3 Theories and Structures

Nodelman and Zalta (2014) have also proposed a form of *ante rem* structuralism that is also not subject to the permutation objection. I will now briefly discuss their view.

Central to Nodelman and Zalta's account is a distinction between two kinds of predication: *exemplification* and *codification*. Exemplification is the form of predication that we are most familiar with. For instance, we have a case of an ordinary object exemplifying a property when we say that Vladimir Putin is the president of Russia. But abstract objects can codify predicates or groups of predicates. For instance, the abstract object redness codifies the property of being red.

According to Nodelman and Zalta (2014, p. 51), mathematical structures are obtained from mathematical theories, along roughly the following lines. There is a canonical way of constructing properties out of propositions via λ-abstraction. A mathematical theory can be identified with the collection of mathematical propositions that it logically entails. Therefore, a mathematical theory is associated with the collection of properties constructed out of the propositions that it entails. And there will be a unique abstract object that encodes exactly those properties. So we may *identify* a mathematical structure with this abstract entity.

This is an *ante rem* form of structuralism that differs from the accounts that we have discussed so far. It associates a unique *ante rem* structure even with each algebraic theory, and thus comes closer to the account that I propose than to Shapiro's version of non-eliminative structuralism. Moreover, it aims at attributing exactly the right kind of incompleteness to mathematical structures. Consider for instance the theory of dense linear orderings. It will be, on Nodelman and Zalta's account, about an abstract entity (structure) that encodes all the properties of this theory. In particular, it encodes *neither* countability or uncountability, since there are both countable and uncountable dense linear orderings.

On Nodelman and Zalta's view, mathematical structures contain mathematical objects. The objects that a structure contains are extracted from the theory from which the structure is obtained. Roughly, objects correspond to the *terms* that the theory contains. The object corresponding to a term of a theory will again be an abstract object that codifies the properties that the theory attributes to it. Thus the Peano Arithmetic structure will be, in some sense, about all and only the familiar objects $0, 1, 2, \ldots$ And this will have as a consequence that there will be no cross-theory identification of mathematical objects: the number 0 of Peano Arithmetic will be distinct from the number 0 of Real Analysis.[4]

However, since mathematical objects are identified by means of clusters of properties satisfied by the denotation of a term according to a given theory, there is no room in their account for numerically distinct but strongly indistinguishable objects (Nodelman and Zalta 2014, p. 73):

An element *of a structure must be uniquely characterizable in terms of the relations of the structure.*

But this is in a tension with ways in which mathematicians tend to speak and think. Take for instance the theory of countable dense linear orderings without endpoints. Intuitively, one would say that any corresponding structure contains countably many numerically distinct objects that are all indistinguishable from each other. But on Nodelman and Zalta's theory, the structure that this theory describes does not contain *any* objects.

The moral of this is, I think, that there is more to the objects of a structure than the discriminating powers of our best theory about the structure. On

---

[4] So on this point Nodelman and Zalta's account is closer to Resnik (1981) than to Shapiro (1997).

my account, there is a sense in which *even arithmetic* is about a structure containing many mathematical objects (arbitrary numbers) that are mutually highly indistinguishable from each other.

8.4 Comparison with Fine

I now turn to a discussion of the ways in which my account of mathematical structures differs from, and has advantages over, the account of mathematical that Fine (1998) sketches. We will see how our philosophical disagreements about the nature of mathematical structures are inextricably bound up with differences in our conception of arbitrary objects.

8.4.1 Dependence. We have seen in Sections 7.2 and 7.5 how the dependence relation is a fundamental ingredient of Fine's theory of arbitrary objects and of his theory of types. And we have also seen (in Section 7.5) how the distinction between one-way dependence and two-way dependence gives rise to an objection against Fine's account of finite order types, and of the natural number structure.

Fine's dependence relation between arbitrary objects is *not* a fundamental feature of the account of arbitrary objects that I have proposed in Chapter 3. On the account that I favour, *correlation* between arbitrary objects is fundamental. This is, in my view, the most important difference between my theory of arbitrary objects and the one that Fine advocates. We will see in the next chapter that in the theory that I propose, relations of dependence can be *defined* in terms of co-variation.

This is related to another global difference between Fine's theory of arbitrary objects and mine. For the account that I propose, Fine's notion of a *system* of variable objects – not to be confused with an arbitrary system![5] – is not fundamental. Systems of variable objects of kind K are merely substructures of the entire structure of arbitrary objects of kind K.

Let us briefly go back to Fine's theory of the order type 2 (Section 7.4). He holds that there is a difference between

1 the system S_1 consisting of the independent variable object p, ranging over all ordered pairs, and the dependent variable objects v_1^2 and v_2^2 that always select, respectively, the first and the second element of each value of p, and

---

[5] See Section 6.1.

2 The system S_2, which is just like S_1, except that in S_2, all variable objects co-depend on each other.

From the perspective of the theory that I propose, given the identity criterion that is captured by Thesis 3.3 (Section 3.4), there is no difference between the systems S_1 and S_2, for their co-variation patterns are identical. Indeed, I cannot see any difference between S_1 and S_2 even from a pretheoretical point of view.

So the arbitrariness objection that Fine raises against his own account of the order type 2 (and then dismisses), cannot be raised against the account of the order type 2 that I propose. On my account, the order type 2 is less complicated than on Fine's account: it is simply identified with what Fine calls the *prototype*, i.e. the generic system S that can be in the state of being any ordered pair (and in no other state).

We have seen that Fine (2017b, Section 5) locates questions of ontological reduction in foundational metaphysics. In Chapter 2, I proposed to scrap foundational metaphysics. But Fine's theory and mine are two rival naive metaphysical accounts of arbitrary objects. Fine recognises a *primitive* relation of dependence between arbitrary objects, whereas I do not. In Section 2.4 I carved the distinction between naive and foundational metaphysics along slightly different lines than Fine does. In particular, I argued that there is room *within* naive metaphysics for debates about reductionism. Here we have a case in point: Fine and I are having a discussion about ontological reductionism.

I agree with Fine that dependence is a fundamental property of arbitrary objects. We disagree, however, about the properties of this dependence relation. I take them to *supervene* on covariation patterns between arbitrary objects; Fine does not. I take the covariation relation between two arbitrary objects to *be* a dependence relation between them. And I believe that a specific dependence relation that is fairly close to what Fine had in mind can be defined in terms of covariation (Section 9.4). Moreover, we will see in Chapter 10 that covariation patterns also allow natural *probabilistic* dependence between arbitrary objects to be defined.

8.4.2 Identity. We have seen that given Fine's criterion of identity for systems of variable objects (Section 7.3), there is *exactly one* generic ω-sequence.

I have sought to qualify this statement. It is a matter of perspective. From within arithmetic, Fine's criterion of identity holds. But from a higher-order perspective, there are many generic ω-sequences that are correlated with each other in complicated ways. I take this perspective-relativity to be in accordance with lived mathematical experience. From the internal perspective of arithmetic, there is only the structure of the natural numbers. But this does not prevent there being many ω-sequences from a higher perspective. There is nothing wrong when a mathematician who says (in an analysis lecture, perhaps):

Example 8.4.

Let A be any ω-sequence.

...

Let B be any ω-sequence that is different from A.

...

The slogan is that in mathematics, you can take multiple copies of everything.

Recall from Section 3.4 that on my account, systems that are in the same state in every possible situation are numerically identical. I take this to be an *absolute* criterion of identity for generic systems.

The criterion that I propose is much weaker than the identity criterion that Fine advocates. For one thing, it is compatible with there being many generic ω-sequences.

In Section 3.6 we saw that what my criterion of identity for arbitrary objects presupposes – i.e. that there are no two situations consisting of exactly the same objects standing in exactly the same relations to each other – is not self-evident. Nevertheless I believe that this presupposition follows from the nature of the modality (afthairetic possibility) that is at work here. Suppose that in probability theory a single loaded die is considered. Say, for instance, that the die is heavily loaded toward coming up 4. Then what is *not* done is to take there to be more than one situations in which the die comes up 4, where a situation here is nothing above and beyond the state of the die (since that is the only object under consideration). Instead, this state is given more *weigth* than the other states of the die.

8.4.3 Self-Instantiation. Non-eliminative structuralism claims that mathematical structures are entities that can be instantiated by systems. So in *some* sense, in Chapter 6 I can be said to propose a form of non-eliminative structuralism: *generic structuralism*, you might say (despite my reservations about this term). On my view, structures are arbitrary objects that can be in the state of being a specific system. So *being in a state* is my proposed explication of the instantiation relation. Nonetheless, most forms of non-eliminative structuralism make additional claims: foundational claims. For instance, many of them claim that mathematical structures constitute the subject matter of mathematics. Such theses go beyond what I want to commit myself to. It is or this reason I rejected the label in the beginning of this chapter (Section 8.1), and am unwilling to commit myself to any -ism.

Fine (1998, Section VI) makes a structuralist proposal without really endorsing it. The sketchiness of his proposal makes it difficult for me to evaluate it. Fine's account of mathematical structures differs from mine in that he does not take a mathematical structure to be a variable object. His aim appears to be to give an account of what it means to be a *preferred system* (Fine 1998, pp. 629–630):

> *from among all of the models that make the theory true, it is often supposed that there is a particular model – the* intended model – *which the theory is properly about. According to the representational structuralist, what distinguishes this model from all of the others is its representational role. The model somehow serves to represent all of the other models, with each element or relation of the model somehow serving to represent the elements and relations of the other models ... It is here that our account of Cantorian abstraction may be of some help.*

Fine's account of the intended model of arithmetic was described in Section 7.5: let us call this intended model F. Now one may ask whether this 'intended' model can or should be *identified* with the natural number structure. If so, then we have a form of non-eliminative structuralism. It is tempting to suppose that Fine intends to be thus understood, given that on his abstractionist account of order types, F *is* the order type ω. Nevertheless, it is not straightworward to see how this works.

First, as was pointed out at the end of Section 7.5, it is natural to wonder whether, on the non-eliminativist reading, Fine's account is vulnerable to Hellman's permutation objection: consider, for instance, the model F^* obtained by 'swapping' the first and second element of F. I suspect that Fine would reply that F^* is not the natural number structure. The reason is that the variable object that it takes to be the number 0 is the variable object that takes the *second* value in every ω-sequence. I will leave it open here whether that is a convincing reply.

Second, Fine's intended model F is not a variable object. So, if instantiation is understood as taking a value, then F is not the kind of entity that can be instantiated.

So it seems that Fine has articulated an eliminative flavour of structuralism *with a preferred system* rather than a form of non-eliminative structuralism. But if this is the right way to interpret Fine's proposal, then, as we saw in Section 5.4.3, the thesis that the mathematical number structure contains mathematical objects (the natural numbers) has to be abandoned.

The account that I propose is simpler. It stops at what Fine calls the *prototype* and takes that to be the structure. Rather than seeing the prototype not being a particular system as a defect, as Fine does (Fine 1998, pp. 618–619), I regard this as a strength of the account that I propose, for it immunises the account against the permutation objection.

8.5 Quantification and Reference

As we saw in Sections 7.3.2 and 7.6, Fine was in his earlier work on arbitrary objects mostly interested in situations where a finite number of arbitrary objects are gradually introduced in a context (in a proof, for instance). He investigated in detail how we *reason* with arbitrary objects thus introduced in particular contexts (Fine 1985a, 1985b).

We have seen in Section 7.3.1 how Fine holds that for a given kind F, there is a *unique* arbitrary F. Since the uniqueness condition is satisfied, we can coin a name a for the independent arbitrary F. Many dependent arbitrary Fs can then be uniquely identified by means of descriptions involving the name a, so they can be given names also.

In this book I am not much concerned concerned with the question how arbitrary objects are introduced in a conversational context and used in everyday mathematical and non-mathematical reasoning. I hold that, for a given kind F, there are multiple arbitrary objects. I take *all* the arbitrary Fs as given all at once, and try to investigate their structure and their properties.

A question that arises for me, but not for Fine, is how we can *refer* to independent arbitrary Fs, given that – for me, at least – there are many. Frege thought that this is an insuperable problem. As was touched upon earlier, he had this to say about taking a mathematician's speech at face value when she says 'take an arbitrary number x and an arbitrary number y' (Frege 1960, p. 109):

This way of speaking is certainly employed; but these letters are not proper names of variable numbers in the way that '2' and '3' are proper names of constant numbers; for the numbers 2 and 3 differ in a specified way, but what is the difference between the variables that are said to be designated by 'x' and 'y'? We cannot say. We cannot specify what properties x has and what differing properties y has. If we associate anything with these letters at all, it is the same vague image for both of them. When apparent differences do show themselves, it is a matter of application; but we are not here talking about these. Since we cannot conceive of each variable in its individual being, we cannot attach any proper names to variables.

This problem is particularly acute since it is fairly generally accepted that *facts about reference are determined by facts about language use.* Breckenridge and Breckenridge and Magidor (2012, p. 380) phrase this thesis as follows:

It is standardly accepted that semantic facts are not primitive: rather, they are determined by use facts (broadly construed) ... Semantic facts supervene on use facts.

So suppose a mathematician says 'Let *x* be an arbitrary number.' And suppose also that there are at least two independent arbitrary numbers *a* and *b*. Then it is difficult to see what we as language users can possibly do to ensure that the expression '*x*' refers to *a* rather than *b* or vice versa.

A version of this problem arises also for Breckenridge and Magidor's (2012, p. 378) semantic theory of arbitrary object terms, which locates the arbitrariness in the reference relation rather than on the side of the objects. In fact, since, as we have seen in Section 3.1, Frege suggested that '*x*' indeterminately refers to a determinate number, this is a problem for Frege himself! If '*x*' refers to some *specific* number, then it is difficult to see what we as language users can possibly do to ensure that it refers to one specific natural number rather than another. In response to this problem, Breckenridge and Magidor *reject* the thesis that reference facts supervene on use facts (Breckenridge and Magidor 2012, Objection 3, p. 380; Kearns and Magidor 2008). This response is also open to me. But the thesis that reference facts supervene on use seems plausible. So I do not think that this is a good way out.

A different response is to ascend to a second-order level and say that '*x*' refers to *the arbitrary* arbitrary *F*. But this does not give us names for arbitrary *F*'s. Moreover, it generates an additional problem at the second-order level: the pressure to recognise multiple independent arbitrary *F*s is just as strong on the second-order level as on the first-order level.

So '*x*' does not name individual independent arbitrary *F*s. This raises the question whether individual independent arbitrary *F*s can be named at all. Dependent arbitrary *F*s can only be named using names for independent arbitrary *F*s. So if the latter cannot be named, then the former cannot be named, either.

Indeed, I think that our mathematician cannot coin a *name* for an arbitrary number. The expression '*x*' is not a name but a *free variable* that is implicitly universally quantified over in a context. Some expressions referring to arbitrary objects are even designed to be re-used in new context. Expressions such as 'Jane Doe' or 'the man on the Clapham omnibus' function as a *dedicated variable* as described in Section 5.4.3.

So there is something deeply right about the theory described in Pettigrew (2008), and the way in which my position concerning the semantics of these expressions differs from Pettigrew's position is somewhat subtle. We agree that they function as dedicated variables, and disagree with Fine who takes them to function as names. But Pettigrew holds that these designated variables range over specific entities, whereas I hold that it is an equally good hypothesis that they range over arbitrary entities.

9 | Reasoning about Generic ω-Sequences

One of the aims of this book is to acquire deeper insight into the nature of mathematical structures. If mathematical structures are generic systems, and if our set-theoretic way of modelling them represents the salient features of generic systems well, then deeper insight into the nature of mathematical structures results from investigating the properties of the proposed set-theoretic representations of generic systems.

This is the subject matter of the present chapter, which is more logical in character than the previous ones. The focus is on one particular mathematical structure: the natural number structure. I take the perspective that is internal to arithmetic (see Section 6.4), from which arithmetic investigates *one* structure.

In Chapter 6 two conceptions of the natural number structure as a generic system were discussed: the full generic ω-sequence and the computable generic ω-sequence. Both of these were conceived as generic systems. The present chapter is devoted to 'reading off' properties of these generic systems from their set-theoretic representations.

I will argue that from a logical point of view, the computable generic ω-sequence $\mathbf{N}_C$ is a more natural model than the full generic ω-sequence $\mathbf{N}$. It is much simpler, while displaying roughly (but not quite) the same logical behaviour. In other words, from a logical point of view, the full generic ω-sequence carries superfluous structure.

9.1 Individual Concepts and Carnapian Modal Logic

A good formal framework is needed for reasoning about arbitrary systems in general, and about generic systems in particular. We want a framework that is conceptually clear, suitable for the intended application, expressively powerful, and easy to use (i.e. not unwieldy or too cumbersome).[1]

---

[1] These requirements are similar to the list of desiderata for a good framework for intensional logic that is given in Belnap and Müller (2014, p. 394).

We want to quantify over the objects that belong to a generic ω-sequence. So a quantificational framework is needed. We will work in a *first-order setting*. Quantification and identity are intimately related notions, so we will have an identity relation in the framework.

Liar-like or set theoretic paradoxes are sometimes taken as a motivation to work in a framework of non-classical logic. I have argued in Section 3.5 that in arbitrary object theory, Russell-like paradoxes loom only if we allow arbitrary objects whose value range includes all higher-order arbitrary objects (over a set of non-arbitrary objects). But I am not aiming for this kind of universality in this chapter or even in this book. So Russellian concerns constitute no reason for me to depart from a classical framework. Some hold that an adequate resolution of the issues surrounding the principle of generic attribution (Thesis 7.4, Section 7.3.3) requires a non-classical framework. Santambrogio, for instance, has formulated an intuitionistic theory of arbitrary objects for this reason (Santambrogio 1987). However, it will be argued in Section 9.6.2 that a satisfactory treatment of comprehension can be obtained in a classical setting. For these reasons, I will stick with classical logic throughout.

A modality (being in a state) is at the heart of both the theory of arbitrary objects and the theory of arbitrary systems. So will work in a framework of *modal logic*. Our set theoretic models of generic ω-sequences make no mention of an accessibility relation on the state space. This means that we are nudged in the direction of S5 modal logic. Instead of representing truths about the generic ω-sequence in a language of modal logic, we could of course quantify directly over states and sets of states. Indeed, this is exactly what we did when we modelled the generic ω-sequence in set theory. But for exactly the same reasons as languages of modal logic have been considered useful in the formal investigation of problems of metaphysics, a language of modal logic is illuminating in the present context. In brief, suitable languages of modal logic allow us to express in a natural way what we want to express, without containing excessive formal structure.

The elements of generic ω-sequences are specific and arbitrary numbers. Both of these are arbitrary objects, and arbitrary objects are *individual concepts* (functions from states to objects). So it appears that we need to work in a framework for reasoning about individual concepts. Carnap's (1956) framework for quantified modal logic fits the bill: it is a calculus of individual concepts. Indeed, Kripke (1992, p. 73) suggested that Fine's theory of arbitrary objects can be formalised in Carnapian quantified modal logic:

Fine has argued in favor of 'arbitrary objects'. Much of what he proposes to treat with his arbitrary objects . . . could be done quite naturally [in Carnapian quantified modal logic] (where . . . at least an intensional predicate for rigidity [is] helpful). Indeed, mathematically arbitrary objects are simply [individual concepts] of a special kind.

Kripke especially had Fine's suggestion of applying arbitrary object theory to elementary Calculus and Analysis in mind (see Section 4.6). But in the present chapter, Kripke's suggestion is applied to generic ω-sequences and is worked out in some detail. The right formal framework for investigating generic ω-sequences is Carnapian quantified modal logic with identity, plus, as Kripke suggests, one distinguished intensional predicate.

Carnapian quantified modal logic is concerned with the modal profiles of individual concepts. These modal profiles are determined by the way in which their *extensions* are correlated with each other in different possible states. Extensions of individual concepts are themselves not individual concepts but objects. So the formalism must somehow be able to talk both about individual concepts *and* about objects. The Carnapian framework succeeds in doing this in a particularly elegant and economical way, by letting the individual variables play a double role – Kripke (1992, p. 72) calls this *Carnapian doublethink*.

Nonetheless, the proof of the pudding is in the eating. I leave it to the reader to judge whether this framework meets the desiderata that are listed at the beginning of this section. I will confine myself here to discussing briefly why other approaches do not meet these desiderata.

The standard framework for quantified modal logic is the *Kripkean framework* (Kripke 1963). It differs from Carnapian modal logic in its intended subject matter. Kripkean quantified modal logic aims at describing the modal properties of *specific objects*, not of individual concepts. It is a very perspicuous and versatile framework. The only problem is that it is not suited for our intended application.

We have seen in Section 6.9 that a coherent conception of higher-order arbitrary objects can be formed. So one might ask whether we should use a framework of *higher-order* Carnapian modal logic, such as that of Bressan (1972). However, such a framework is notoriously cumbersome to work in (Belnap and Müller 2014, p. 394). Moreover, for our purposes, it contains excess structure. I will not be interested in the higher-order properties of generic ω-sequences, and we will see that first-order Carnapian quantified modal logic already has a high degree of expressive power.

We have seen that Fine regards the *dependence* relation between arbitrary objects as a fundamental and irreducible feature of his metaphysical theory. The dependence relation also plays a fundamental role in his formal framework for reasoning with arbitrary objects. It allows him to express functional relations between variables in his interpreted formal languages. Indeed, we saw in Section 4.6 that Fine expressed hopes that the theory of arbitrary objects may provide a good framework for the formal investigation of Skolem functions. This indicates that there is a close relationship between Fine's formal framework, on the one hand, and the framework of Dependence Logic (as discussed in Väänänen 2007, for instance), on the other hand. However, I have argued in Section 7.2 that dependence should not be taken to be a fundamental notion in the theory of arbitrary objects – not in the particular way that Fine does, anyway. Moreover, we will see that the framework of Dependence Logic, or, pretty much equivalently, Fine's framework, is not sufficiently expressive for our purposes.

In what follows, I propose a Carnapian formal framework, and discuss some of the key properties of the generic and the computational ω-sequence that can be expressed in this framework. Proofs of the key results will not be given here: the reader who is interested in proofs can find them in Horsten and Speranski (2018).

9.2 Generic Truth

Let's now get down to business. In this section, the Carnapian formal framework is described in some detail.

9.2.1 Revisiting the full and the computable generic ω-sequence. Without loss of generality, and purely for purposes of exposition, we may identify the underlying countably infinite plurality of objects with $\mathbb{N}$ itself. Let

$$\mathbb{G} := \text{the collection of all permutations of } \mathbb{N}.$$

The set $\mathbb{G}$ models the *full generic ω-sequence*. For all intents and purposes, it is equivalent to the set-theoretic way of modelling the full generic ω-sequence **N** in Section 6.4. Clearly $|\mathbb{G}| = 2^{\omega}$, so there are continuum many states that the full generic ω-sequence can be in.

Take σ to be the signature of Peano Arithmetic, i.e. $\{0, \mathsf{s}, +, \times, =\}$, and $\mathfrak{N}$ to be its *standard model*, which serves as the intended interpretation of

σ. Naturally, every $\pi \in \mathbb{G}$ induces its own isomorphic copy $\pi [\mathfrak{N}]$ of $\mathfrak{N}$, given by

$$\pi [\mathfrak{N}] \models \pi (i) = 0 \qquad\qquad \Longleftrightarrow \qquad \mathfrak{N} \models i = 0;$$

$$\pi [\mathfrak{N}] \models \mathsf{s} (\pi (i)) = \pi (j) \qquad \Longleftrightarrow \qquad \mathfrak{N} \models \mathsf{s} (i) = j;$$

$$\pi [\mathfrak{N}] \models \pi (i) + \pi (j) = \pi (k) \qquad \Longleftrightarrow \qquad \mathfrak{N} \models i + j = k;$$

$$\pi [\mathfrak{N}] \models \pi (i) \times \pi (j) = \pi (k) \qquad \Longleftrightarrow \qquad \mathfrak{N} \models i \times j = k.$$

In other words, facts about $\pi [\mathfrak{N}]$ can be obtained by applying π^{-1} to the intended interpretations of the symbols in σ.

An *arbitrary number* in $\mathbb{G}$ is an individual concept that for each state π picks out some position in π, regarded as an ω-sequence. Formally, arbitrary numbers in $\mathbb{G}$ can be identified with functions from $\mathbb{G}$ to $\mathbb{N}$. Denote by $\mathbb{A}$ the collection of all arbitrary numbers in $\mathbb{G}$. Evidently, $|\mathbb{A}| = 2^{2^{\omega}}$.

The *specific numbers* are treated as limiting cases of arbitrary numbers. To see how this works, consider the arbitrary number α such that for every $\pi \in \mathbb{G}$,

$$\alpha (\pi) \ = \ \pi (0).$$

For each state $\pi \in \mathbb{G}$, α picks out the object that plays the role of 0 at π (regarded as an ω-sequence). This α may be thought of as the *specific number* 0 in $\mathbb{G}$. Then, in general, we define the *specific number n* in $\mathbb{G}$ to be the arbitrary number α given by

$$\alpha (\pi) \ = \ \pi (n).$$

Let $\mathbb{S}$ be the collection of all specific numbers. Obviously, $|\mathbb{S}| = \omega$.

Of course most of the permutations of $\mathbb{N}$ induce non-computable presentations. In the spirit of computational structuralism (Benacerraf 1965, pp. 275–277; Halbach and Horsten 2006; Horsten 2012), one might insist that the structures that instantiate our intended interpretation must represent *computable* notation systems.

This line of reasoning leads to narrowing the state space $\mathbb{G}$ to

$$\mathbb{G}^c := \text{the collection of all computable permutations of } \mathbb{N},$$

where we may regard its elements as constructive systems of notations. This yields a computable isomorphism type. $\mathbb{G}^c$ models the *computable generic ω-sequence*. For all intents and purposes, $\mathbb{G}^c$ is equivalent to the computable generic ω-sequence $\mathbf{N}_C$ of Section 6.6. Evidently there are only countably many Turing machines, hence $|\mathbb{G}^c| = \omega$.

As in the case of the full generic ω-sequence, we can then go on to identify the collection of all arbitrary numbers in $\mathbb{G}^c$ with the set $\mathbb{A}^c$ of all functions from $\mathbb{G}^c$ to $\mathbb{N}$. Clearly, $|\mathbb{A}^c| = 2^\omega$. As in $\mathbb{G}$, specific numbers are treated as limiting values of arbitrary numbers.

We shall be concerned with both $\mathbb{G}$ and $\mathbb{G}^c$. I shall often write **G** when there is no need to differentiate between these two.[2] Similarly, let **A** be the collection of all functions from **G** to $\mathbb{N}$, and let S be the collection of all functions from **G** to $\mathbb{N}$ such that $\alpha\,(\pi)\;=\;\pi\,(n)$.

9.2.2 Language and syntax. Let us now turn to the description of the formal first-order modal language $\mathcal{L}$ that we shall be working with.

The language of first-order Peano Arithmetic, $\mathcal{L}_1$, with signature $\sigma_1 = \{0, s, +, \times\}$, is sufficiently familiar. In addition to the symbols of the signature σ of Peano Arithmetic, our formal language $\mathcal{L}$ includes

- a countable set $\mathrm{Var} = \{x, y, z, \ldots\}$ of *variables*;
- the *connective* symbols $\neg$ and $\vee$;
- the *quantifier* symbol $\exists$;
- the *modal operator* symbol $\Diamond$, read as 'it is possible that';
- the extra unary predicate symbol Sp, read as '...is specific'.

Notice that in the context of first-order logic, we shall treat $\wedge$, $\rightarrow$, and $\forall$ as defined rather than as primitive.

The $\mathcal{L}$-*formulas* are then built up in the usual manner:

- if t_1 and t_2 are σ-terms, then $t_1 = t_2$ is an $\mathcal{L}$-formula;
- if Φ is an $\mathcal{L}$-formula, then $\neg\Phi$ is an $\mathcal{L}$-formula;
- if Φ and Ψ are $\mathcal{L}$-formulas, then $\Phi \vee \Psi$ is an $\mathcal{L}$-formula;
- if x is a variable and Φ is an $\mathcal{L}$-formula, then $\exists x\,\Phi$ is an $\mathcal{L}$-formula;
- if Φ is an $\mathcal{L}$-formula, then $\Diamond\Phi$ is an $\mathcal{L}$-formula;
- if t is a σ-term, then $\mathsf{Sp}\,(t)$ is an $\mathcal{L}$-formula.

I abbreviate $\neg\,(\neg\Phi \vee \neg\Psi)$ to $\Phi \wedge \Psi$, $\neg\Phi \vee \Psi$ to $\Phi \rightarrow \Psi$, $\neg\exists x \neg\Phi$ to $\forall x\,\Phi$, and $\neg\Diamond\neg\Phi$ to $\Box\Phi$ Moreover, I shall often abbreviate $\Phi \rightarrow \Psi \wedge \Psi \rightarrow \Phi$ further to $\Phi \leftrightarrow \Psi$. Of course, x, y, z, ... are intended to range over arbitrary natural numbers. In the sequel, restricted quantification over *specific* numbers will play a role. For this reason, I will sometimes abbreviate $\forall x(\mathsf{Sp}(x) \rightarrow \ldots)$ as $\forall^{\mathsf{Sp}}x(\ldots)$. Moreover, for variables that are so restricted, I will occasionally use $n, m, \ldots$ rather than $x, y, \ldots$

---

[2] By using this notation, I am then also not differentiating between the generic ω-sequence (see Section 6.7) and the set-theoretic structures that are intended to model it. This should not cause confusion: the present chapter is about set-theoretic ways of modelling the generic ω-sequence.

As discussed in Section 10.2, the $\mathcal{L}$-formula $x = y$ will not express the *identity* of x and y, but only their 'coincidence' at a given possible world. Still, real identity turns out to be expressible, via

$$\simeq (x, y) := \square (x = y),$$

as expected. Observe that Sp is the only symbol of $\mathcal{L}$ with *intensional* meaning, and its presence will play a crucial role in the expressive power of $\mathcal{L}$. Note also that the first-order σ-formulas form a subset of the $\mathcal{L}$-formulas. These formulas will occasionally be called *purely arithmetical*.

9.2.3 Interpretation. Any *state* or *possible world* is a permutation of $\mathbb{N}$, which may or may not be computable, depending on what we take $\mathbf{G}$ to be. As mentioned earlier, we associate with each $\pi \in \mathbf{G}$ the σ-structure $\pi [\mathbb{N}]$. More precisely,

$$\pi [\mathbb{N}] := \langle \mathbb{N}, 0^{\pi}, \mathsf{s}^{\pi}, +^{\pi}, \times^{\pi}, =^{\pi} \rangle,$$

where $=^{\pi}$ is the ordinary equality relation on $\mathbb{N}$, and the others are given by

$$0^{\pi} := \pi (0);$$
$$\mathsf{s}^{\pi} (i) := \pi \left(\pi^{-1} (i) + 1 \right);$$
$$i +^{\pi} j := \pi \left(\pi^{-1} (i) + \pi^{-1} (j) \right);$$
$$i \times^{\pi} j := \pi \left(\pi^{-1} (i) \times \pi^{-1} (j) \right).$$

Here 0, s, $+$, and $\times$ on the right sides have their standard meaning, as in $\mathbb{N}$. Thus, for instance, we define $i +^{\pi} j$ to be (roughly speaking) the number that plays at π the role of the sum of the role that i plays at π and the role that j plays at π.

By a *valuation* in $\mathbf{A}$, we mean simply a function from Var to $\mathbf{A}$. Naturally, every valuation γ in $\mathbf{A}$ can be inductively extended to σ-terms:

$$\gamma (0) := \lambda \pi . [0^{\pi}];$$
$$\gamma (\mathsf{s} (t)) := \lambda \pi . [\mathsf{s}^{\pi} ((\gamma (x)) (\pi))];$$
$$\gamma (t_1 + t_2) := \lambda \pi . [(\gamma (t_1)) (\pi) +_{\pi} (\gamma (t_2)) (\pi)];$$
$$\gamma (t_1 \times t_2) := \lambda \pi . [(\gamma (t_1)) (\pi) \times_{\pi} (\gamma (t_2)) (\pi)].$$

Here $\lambda \pi . [\ldots \pi \ldots]$ denotes the function which maps each π in $\mathbf{G}$ to $\ldots \pi \ldots$.

Of course the 'standard' interpretation of Sp in $\mathbf{G}$ is

$$S := \{ \lambda \pi . [\pi (n)] \mid n \in \mathbb{N} \}.$$

Nevertheless, it is useful in some contexts to allow Sp to be interpreted by other subsets of A as well. Given an $S \subseteq A$, we define, for any $\mathcal{L}$-formula Φ, valuation γ in A and world $\pi \in G$, what it means for Φ to be *true in* $\langle G, S \rangle$ *at* π *under* γ, written $\langle G, S \rangle \models_\pi \Phi [\gamma]$, as follows:

- $\langle G, S \rangle \models_\pi t_1 = t_2 [\gamma]$ iff $(\gamma(t_1))(\pi) = (\gamma(t_2))(\pi)$;
- $\langle G, S \rangle \models_\pi \neg \Phi [\gamma]$ iff $\langle G, S \rangle \not\models_\pi \Phi [\gamma]$;
- $\langle G, S \rangle \models_\pi \Phi \vee \Psi [\gamma]$ iff $\langle G, S \rangle \models_\pi \Phi [\gamma]$ or $\langle G, S \rangle \models_\pi \Psi [\gamma]$;
- $\langle G, S \rangle \models_\pi \exists x \Phi [\gamma]$ iff there exists $\alpha \in A$ such that $\langle G, S \rangle \models_\pi \Phi [\gamma_\alpha^x]$;
- $\langle G, S \rangle \models_\pi \Diamond \Phi [\gamma]$ iff there exists $\pi' \in G$ such that $\langle G, S \rangle \models_{\pi'} \Phi [\gamma]$;
- $\langle G, S \rangle \models_\pi \mathsf{Sp}(t) [\gamma]$ iff $\gamma(t) \in S$.

Here we use γ_α^x for the valuation which agrees with γ, except that $\gamma_\alpha^x(x) = \alpha$, viz.

$$\gamma_\alpha^x(y) := \begin{cases} \gamma(y) & \text{if } y \neq x, \\ \alpha & \text{if } y = x. \end{cases}$$

Clearly if Φ is of the form $\Phi(x_1, \ldots, x_\ell)$, i.e. the free variables of Φ are among $x_1, \ldots, x_\ell$, then it is immaterial what values γ assigns to the elements of $\text{Var} \setminus \{x_1, \ldots, x_\ell\}$, so we may write

$$\langle G, S \rangle \models_\pi \Phi [\gamma(x_1), \ldots, \gamma(x_\ell)],$$

or more explicitly $\langle G, S \rangle \models_\pi \Phi [x_1/\gamma(x_1), \ldots, x_\ell/\gamma(x_\ell)]$. Furthermore, when Φ is an $\mathcal{L}$-sentence, i.e. no variable occurs free in Φ, this becomes $\langle G, S \rangle \models_\pi \Phi$.

Proposition 9.2.1. Let $S \subseteq A$ and $\pi \in G$. For any purely arithmetical $\mathcal{L}$-formula $\Phi(x_1, \ldots, x_\ell)$ and $(\alpha_1, \ldots, \alpha_\ell) \in A^\ell$,

$$\langle G, S \rangle \models_\pi \Phi [\alpha_1, \ldots, \alpha_\ell] \iff \pi[\mathbb{N}] \models \Phi [\alpha_1(\pi), \ldots; \alpha_\ell(\pi)].$$

Corollary 9.2.2. The collection of all purely arithmetical sentences true in $\langle G, S \rangle$ at π coincides with the first-order theory of $\mathbb{N}$.

We say Φ is *generically true in* $\langle G, S \rangle$ *under* γ, written $\langle G, S \rangle \models \Phi [\gamma]$, iff $\langle G, S \rangle \models_\pi \Phi [\gamma]$ for all $\pi \in G$. As before, in the case where Φ is an $\mathcal{L}$-sentence we can omit γ and write $\langle G, S \rangle \models \Phi$: in this case we will call Φ a *generic truth* (of G).

Definition 9.1. Let $Tr_\mathcal{L}(G)$ be the collection of all generic truths of G that are expressible in $\mathcal{L}$.

Note that the notion of generic truth does not presuppose there being an 'actual' world in which arbitrary numbers take their 'actual' values.

Corollary 9.2.3. The collection of all purely arithmetical sentences that are generically true in $\langle \mathbf{G}, S \rangle$ coincides with the first-order theory of $\mathbb{N}$.

A set Γ of $\mathcal{L}$-formulas whose free variables are among $x_1, \ldots, x_k$ is called *satisfiable* if there is a k-tuple $\vec{\alpha} \in \mathbf{A}^k$ such that $\mathbf{G} \models \Phi(\vec{\alpha})$ for all $\Phi \in \Gamma$, and *finitely satisfiable* if every finite subset of Γ is satisfiable. Not surprising, we have (Horsten and Speranski 2018, Proposition 4.13):

Proposition 9.2.4. There exists a set of $\mathcal{L}$-formulas which is satisfiable, but not finitely satisfiable.

So compactness fails for $\mathcal{L}$ as interpreted in $\mathbf{G}$. This is a first indication that the present framework is expressively stronger than first-order logic.

In Chapter 7, we stated that it would be good to have a setting in which both generic and classical readings can be naturally expressed. It now emerges that the Carnapian framework delivers on this count: the distinction in $\mathcal{L}$ between coincidence ($=$) and real identity ($\simeq$) provides us with the means to express what we want.

As an example, consider the predicate

$$x \text{ is a specific number.}$$

The generic reading of this predicate is expressed simply by the formula

$$\exists^{Sp} y(x = y).$$

As pointed out in Section 7.3.3, this condition trivially holds, even of diagonal numbers, for instance. The classical reading of the predicate is expressed by the formula

$$\exists^{Sp} y(x \simeq y).$$

This is a condition that only holds of specific numbers.

9.2.4 Carnapian Quantified Modal Logic. Observe that since there is no restriction on π' in the defining condition for $\langle \mathbf{G}, S \rangle \models_\pi \Diamond \Phi[\gamma]$, the intended accessibility relation is, in effect, the Cartesian square of $\mathbf{G}$. Therefore $\Diamond$ satisfies the propositional laws of **S5**.

One readily verifies that in $\langle \mathbf{G}, S \rangle$, *Leibniz's scheme of the indiscernibility of identicals*

$$\forall x \forall y \left(x = y \rightarrow \left(\Phi(x) \leftrightarrow \Phi[x/y] \right) \right)$$

fails already for the Sp-free $\mathcal{L}$-formulas. Consider, for instance, *two* arbitrary numbers α and β that coincide in value at some state π. Let

$$\Phi := \Diamond(x \neq y).$$

Then at π, we have $(x = y)[x/\alpha, y/\beta]$. But we do not have, at π, that

$$\left(\Phi \leftrightarrow \Phi\left[y \backslash x\right]\right)[x/\alpha, y/\beta]$$

holds, for $\Diamond(x \neq y)[x/\alpha, y/\beta]$ holds (since α and β differ in value at some state), whereas $\Diamond(y \neq y)[x/\alpha, y/\beta]$ fails.

It is easy to see that the *Barcan formula*

$$\forall x \,\Box\Phi\,(x) \rightarrow \Box\forall x\, \Phi\,(x)$$

and its converse both hold in $\langle \mathbf{G}, S\rangle$ (Heylen 2010, p. 357; Williamson 2013, Sections 2.1–2.2). The so-called *Ghilardi formula*

$$\Box\exists x\, \Phi\,(x) \rightarrow \exists x\,\Box\Phi\,(x),$$

which can be regarded as a *choice principle*, fails already for the Sp-free $\mathcal{L}$-formulas, although its converse holds in $\langle \mathbf{G}, S\rangle$, of course (Heylen 2010, p. 358 n1; Williamson 2013, pp. 54–56). For consider the $\mathcal{L}$-formula

$$\Psi := (x = 0) \wedge \neg\Box\,(x = 0).$$

Then $\langle \mathbf{G}, S\rangle \models \Box\exists x\, \Psi$ and yet $\langle \mathbf{G}, S\rangle \not\models \exists x\,\Box\Psi$.

In the light of this, the following standard axioms and rules inference of Carnapian quantified S5 system $\mathbf{C}$ (see e.g. Heylen 2010, p. 357) can be verified to hold in $\mathbf{G}$:

Axiom 9.2 (Tautologies).

All classical tautologies.

Axiom 9.3 (Barcan).

$$\forall x\,\Box\Phi(x) \rightarrow \Box\forall x\Phi(x).$$

Axiom 9.4 (Self-identity).

$$x = x.$$

Axiom 9.5 (Restricted Leibniz).

$$x = y \rightarrow \left(\Phi\,(x) \leftrightarrow \Phi\left[x\backslash y\right]\right), \text{ for all } \Phi \in \mathcal{L}\backslash\{\Box, \mathsf{Sp}\}.$$

Axiom 9.6 (Universal instantiation).

$$\forall x \Phi \to \Phi[x \backslash y].$$

Axiom 9.7 (Necessitation).

$$\text{If } \mathbf{C} \vdash \Phi, \text{ then } \mathbf{C} \vdash \Box \Phi.$$

Axiom 9.8 (S5).

The S5 axioms of propositional modal logic.

Axiom 9.9 (Modus ponens).

$\mathbf{C}$ is closed under the rule of modus ponens.

Axiom 9.10 (Universal generalisation).

$$\text{If } \mathbf{C} \vdash \Phi \to \Psi, \text{ and } x \text{ is not free in } \Phi, \text{ then } \mathbf{C} \vdash \Phi \to \forall x \Psi.$$

9.3 Definability and Indiscernibility

9.3.1 Specific numbers. Recall from Section 9.2.3 that by a *specific number* we mean a function of the form

$$\lambda \pi . [\pi (n)],$$

where $n \in \mathbb{N}$.

Definition 9.11. Given an $S \subseteq \mathbf{A}$, call $R \subseteq \mathbf{A}^\ell$ *definable* in $\langle \mathbf{G}, S \rangle$ iff there is an $\mathcal{L}$-formula $\Phi(x_1, \ldots, x_\ell)$ such that for every $(\alpha_1, \ldots, \alpha_\ell) \in \mathbf{A}^\ell$,

$$(\alpha_1, \ldots, \alpha_\ell) \in R \iff \langle \mathbf{G}, S \rangle \models \Phi[\alpha_1, \ldots, \alpha_\ell].$$

As usual, we can speak of definability of individual arbitrary numbers. For ease of readability, in this section I drop the qualification '$\langle \mathbf{G}, S \rangle$' to the form of definability that is intended.

The 'coordinatewise' addition and multiplication, viz.

$$R_+ := \{(\alpha_1, \alpha_2, \lambda \pi . [\alpha_1 (\pi) + \alpha_2 (\pi)]) \mid (\alpha_1, \alpha_2) \in \mathbf{A}^2\} \quad \text{and}$$
$$R_\times := \{(\alpha_1, \alpha_2, \lambda \pi . [\alpha_1 (\pi) \times \alpha_2 (\pi)]) \mid (\alpha_1, \alpha_2) \in \mathbf{A}^2\},$$

can then be uniformly defined by

$$\Phi_+ (x, y, z) := x + y \simeq z \quad \text{and} \quad \Phi_\times (x, y, z) := x \times y \simeq z,$$

respectively. Let us temporarily pretend that σ includes neither 0 nor s. Still, the specific numbers $\lambda w.[0]$ and $\lambda w.[1]$ will be uniformly definable by

$$\Phi_0(x) := \forall y\, \Phi_+(y,x,y) \quad \text{and} \quad \Phi_1(x) := \forall y\, \Phi_\times(y,x,y),$$

respectively. Hence the 'coordinatewise' successor function

$$R_S := \{(\alpha, \lambda w.[\alpha(w)+1]) \mid \alpha \in A\}$$

can be uniformly defined by

$$\Phi_S(x,y) := \exists z\left(\Phi_1(z) \wedge \Phi_+(x,z,y)\right).$$

Observe in passing that as far as definability is concerned, there is no significant difference between the signature σ and, say, the smaller signature $\{+, \times, =\}$; in fact, any other σ' would do as well provided that the standard models of σ and σ' are first-order interdefinable.

Now for each $n \in \mathbb{N}$, denote $\lambda w.[n]$ by **n**. Define a sequence $A_0(x)$, $A_1(x), \ldots$ of $\mathcal{L}$-formulas by recursion:

$$A_n(x) := \begin{cases} x \simeq 0 & \text{if } n = 0, \\ \exists y\left(A_{n-1}(y) \wedge x \simeq s(y)\right) & \text{if } n > 0. \end{cases}$$

In the light of this recursive definition of formulae, it is evident that the following holds:

Proposition 9.3.1. For each $n \in \mathbb{N}$, the specific number **n** is uniformly definable in $\mathcal{L}$ by the $\mathcal{L}$-formula $A_n(x)$.

Proposition 9.3.1 shows that the formulas A_n allow us to introduce, by explicit definition, *proper names* **n** for the specific numbers. These proper names still express individual concepts. Nonetheless, they can be called *structurally rigid* proper names, for the name **n** picks out, in each state π, the object that plays the structural role of the number n in that state.

Furthermore, no non-specific number is definable (Horsten and Speranski 2018, Theorem 4.5):

Theorem 9.3.2. For every $\alpha \in A$:

$$\alpha \in S \Leftrightarrow \alpha \text{ is definable in } \mathcal{L}.$$

This suggests that Frege was right when he wrote that 'since we cannot conceive of each variable in its individual being, we cannot attach any proper names to variables [i.e. arbitrary numbers]' (Frege 1984, p. 195).

In particular, the *constant functions* c_n, i.e. the arbitrary numbers $\lambda \pi.[n]$ that pick out the same object in each state, are undefinable. If we had names for constant functions, then they would of course be *rigid designators* in the sense of (Kripke 1980). It is mildly interesting to observe that in the present structuralist context, proper names for natural numbers (as individual concepts) are not Kripkean rigid designators, but instead are what I have called structurally rigid individual concepts.

Even though the *set* of specific numbers is not explicitly definable in $\mathcal{L}$ (as we shall see below), we can *inductively* define it by means of the following three axioms:

Axiom 9.12. $\forall x : x \simeq 0 \rightarrow \mathsf{Sp}(x)$.

Axiom 9.13. $\forall x : \mathsf{Sp}(x) \rightarrow \mathsf{Sp}(x + \underline{1})$.

Axiom 9.14. $\forall x : x \not\simeq 0 \wedge \mathsf{Sp}(x) \rightarrow \exists y : \mathsf{Sp}(y) \wedge y + \underline{1} \simeq x$.

Theorem 9.15. $\langle \mathsf{G}, S \rangle \models$ 9.12–9.14 if and only if S is the set of the specific natural numbers.

So if we assume that 9.12–9.14 are satisfied, we may (and henceforth will) write $\mathsf{G} \models \varphi$ instead of $\langle \mathsf{G}, S \rangle \models \varphi$ (for $\varphi \in \mathcal{L}$) because the interpretation of Sp is now fixed.

9.3.2 Diagonal numbers. Recall that a *diagonal number* is an arbitrary number that takes every value in $\mathbb{N}$ in exactly one state. It is not hard to see that:

Proposition 9.3.3. The property of being a diagonal number is $\mathcal{L}$-definable.

Proof An arbitrary number is a diagonal number if and only if it satisfies the formula

$$D(x) := \forall^{\mathsf{Sp}}n\forall^{\mathsf{Sp}}m\{\Diamond x = n \wedge \forall y[\Diamond(x = n \wedge y = m) \rightarrow \Box(x = n \rightarrow y = m)]\}.$$

☐

We observed in Section 6.4 that since $\mathbb{G}$ contains 2^ω states, it contains no diagonal numbers. Moreover, since $\mathbb{G}^c$ contains ω states, it contains 2^ω diagonal numbers. So, in particular, the principle

Axiom 9.16. $\exists x D(x)$

is true in $\mathbb{G}^c$, but not in $\mathbb{G}$. Therefore:

Corollary 9.3.4. $\mathbb{G}$ and $\mathbb{G}^c$ are not elementary equivalent.

Theorem 9.3.2 of Section 9.3 entails that in particular no single diagonal number is definable in $\mathcal{L}$. In the next section we will see that diagonal numbers play a role in the question of the definability of a well-ordering on the arbitrary numbers.

Clearly axiom 9.16 expresses that there are exactly ω many states. This is the key observation in the proof that

Theorem 9.3.5. The structure $\mathbb{G}^c$ can be categorically axiomatised in $\mathcal{L}$.

On the other hand, it can be shown that (Horsten and Speranski 2018, Corollary 4.11):

Theorem 9.3.6. The structure $\mathbb{G}$ cannot be categorically axiomatised in $\mathcal{L}$.

Again this bears witness to a sense in which $\mathbb{G}$ is less tractable than $\mathbb{G}^c$.

9.3.3 Indiscernibility. In keeping with the spirit of structuralism, we want the elements of the underlying set to be to the highest possible degree (compatible with the presence of identity) indistinguishable from each other. The elements of the underlying set are canonically represented among **A** as the constant functions. So we want the constant functions to be *structural objects*.

Recall the definition of a 2-type (Button and Walsh 2018, p. 362):

Definition 9.17. For any $\{\alpha_1, \alpha_2\} \subseteq \mathbf{A}$, let

$$\text{Type}\,(\alpha_1, \alpha_2) := \left\{ \Phi\,(x, y) \mid \Phi \text{ is an } \mathcal{L}\text{-formula and } \mathbf{G} \models \Phi\,[\alpha_1, \alpha_2] \right\}.$$

Of course this definition can be straightforwardly extended to yield the notion of an n-type for any $n \in \mathbb{N}$.

Definition 9.18. Arbitrary numbers α_1 and α_2 are said to be *relatively distinguishable* if and only if

$$\text{Type}\,(\alpha_1, \alpha_2) \neq \text{Type}\,(\alpha_2, \alpha_1).$$

Then α_1 and α_2 are not relatively distinguishable if $\text{Type}\,(\alpha_1, \alpha_2) = \text{Type}\,(\alpha_2, \alpha_1)$ (Ladyman et al. 2012, p. 171). Pairs that are not relatively discriminable are called *two-indiscernibles* (Button and Walsh 2018, pp. 362–363). Two-indiscernibility is a strong notion for languages that contain the identity symbol.

Let $\mathsf{C} \equiv \{c_1, c_2, c_3, \ldots\}$ be the set of constant functions in **A**. Constant functions c_i and c_j are not relatively distinguishable for any $i, j \in \mathbb{N}$ (Horsten and Speranski 2018, Theorem 4.12):

Theorem 9.3.7. For any $\{i, j\} \subseteq \mathbb{N}$ with $i \neq j$,

$$\mathrm{Type}\,(c_i, c_j) \;=\; \mathrm{Type}\,(c_j, c_i).$$

In fact, more can be said. For this, we need the following definition (Button and Walsh 2018, p. 381):

Definition 9.19. A subset $X \subseteq \mathbf{A}$ with at least n elements is n-indiscernible if and only if for any two n-element sequences of distinct elements $\overline{\alpha} = \langle \alpha_1, \alpha_2, \ldots, \alpha_n \rangle$ and $\overline{\beta} = \langle \beta_1, \beta_2, \ldots, \beta_n \rangle$ from X, we have

$$\mathrm{Type}(\overline{\alpha}) = \mathrm{Type}(\overline{\beta}).$$

When X has infinitely many elements, we say that X is ω-indiscernible if and only if X is n-indiscernible for each $n \geq 1$.

Then an induction on the complexity of formulae shows that:

Theorem 9.20. $\mathbf{C}$ is ω-indiscernible.

Intuitively, while Theorem 9.3.2 guarantees that all and only the specific numbers in $\mathbf{G}$ are maximally distinguishable from each other, Theorems 9.3.7 and 9.20 tell us that the constant functions lie at the opposite extreme of the spectrum.

Notions of indiscernibility have an importance in the theory of arbitrary mathematical entities that it scarcely has in other contexts.

Consider the case of set theory. Suppose we have two sets x and y that are 2-indiscernibles for the standard first-order language of set theory. Then the possibility that we might some day have move to some language extension in which x and y have become 2-discernible cannot be excluded.

Now consider two arbitrary natural numbers c_1 and c_2 that are 2-indiscernibles for our language $\mathcal{L}$ – in particular, let us assume that that $c_1, c_2 \in \mathbf{C}$. If the underlying set of our set of arbitrary natural numbers were taken to be collection of physical entities, as in Section 6.3, then we could distinguish c_1 and c_2 in a language extension where we had *names* for these physical entities. We could simply say that c_1 is the arbitrary natural number that always takes the value v, whereas c_2 is the arbitrary natural number that always takes the value w.

But it is hard to see how we could ever have a *mathematical* language in which c_1 and c_2 are 2-discernible. The identity conditions of c_1 and c_2 are given by the *role* they play in each state. For instance, in some state π, the arbitrary number c_1 may play the role of being the third natural number. But the states cannot be identified *independently* from the mathematical roles that the underlying objects play in them. I conclude from this that notions of indiscernibility are more *absolute* in the theory of arbitrary numbers than they are in set theory, for instance.

9.4 Determination and Independence

In this section I want to take a few first steps in the direction of isolating and investigating important relations between arbitrary numbers and hierarchies that these concepts give rise to.

9.4.1 Determination. Dependence Logic (see Väänänen 2007) investigates the following notion of dependence (Väänänen and Grädel 2013, p. 399):

We ... focus on the strongest notion of dependence, namely functional dependence. This is the kind of dependence in which some given variables absolutely deterministically determine some variables, as surely as x and y determine x+y and x · y in elementary arithmetic.

This notion of determination can be applied to arbitrary natural numbers. For simplicity, let us concentrate on the concept of *one* arbitrary natural number determining another arbitrary number. We can define this as follows:

Definition 9.21. An arbitrary number α is *determined* by an arbitrary number β (denoted as $\alpha \preceq \beta$) if and only if for every specific number m, there is a specific number n such that whenever β takes value m, then α takes value n.

This can be formulated in $\mathcal{L}$ as

$$\mathsf{Det}(x, y) \equiv \forall^{\mathsf{Sp}} m \exists^{\mathsf{Sp}} n \square (y = m \rightarrow x = n).$$

In other words, for all $\alpha, \beta \in \mathsf{A}$: α determines β according to G if and only if

$$\mathsf{G} \models \mathsf{Det}(x, y)[x/\alpha, y/\beta].$$

As we saw in Section 3.9, a nineteenth-century Analyst would express this by saying that 'the variable quantity α is a *function* of the variable quantity β'.

Observe that this does not quite coincide with Fine's notion of dependence between arbitrary numbers. We have seen in Section 7.2 that on Fine's view, there can be two pairs of arbitrary numbers $\langle \alpha, \beta \rangle$ and $\langle \alpha', \beta' \rangle$ that are modally correlated with each other in exactly the same way but that differ in dependence (e.g. where β depends on α but β' does not depend on α'). Evidently this is not possible on the notion of determination that is defined here.

The notion of determination is related to partitions induced by arbitrary natural numbers. Every $\alpha \in$ A induces a *partition* of G:

Proposition 9.4.1. For all $\alpha \in$ A: $\{S \subseteq \mathbb{N} \mid S = \alpha^{-1}(n)$ for some $n \in \mathbb{N}\}$ is a partition of G.

For every $\alpha \in$ A, let $\mathcal{P}_\alpha$ denote the partition of G induced by α. Then we see that *partition refinement* generates a criterion for determination:

Proposition 9.4.2. For all $\alpha, \beta \in$ A: α determines β according to G if and only if $\mathcal{P}_\alpha$ is a partition refinement of $\mathcal{P}_\beta$.

The following are elementary properties of the determination relation:

Proposition 9.22. The determination relation $\preceq$ is

1 Reflexive;
2 Transitive;
3 Not symmetric.

Moreover, it is also easy to see that there are arbitrary numbers α, β such that $\alpha \not\preceq \beta$ and $\beta \not\preceq \alpha$.

This means that the notion of determination generates a *degree structure*:

Definition 9.23. For every $\alpha \in$ A:

$$\mathsf{d}(\alpha) \equiv \{\beta \in \mathsf{A} : \alpha \preceq \beta \text{ and } \beta \preceq \alpha\}.$$

Then the relation $\preceq$ can of course be lifted from arbitrary natural numbers to their degrees and is seen to be a partial ordering relation. Call the resulting fine-grained degree structure $\mathsf{D}(\mathsf{G})$.

At this point the distinction between $\mathbb{G}$ and $\mathbb{G}^c$ is significant.

Proposition 9.24. If δ is a diagonal number in G, then $\alpha \preceq \delta$ for every arbitrary number α in G.

We have seen that $\mathbb{G}^c$ contains diagonal numbers (and $\mathbb{G}$ does not). So the degree $\mathsf{d}(\delta)$ of the diagonal numbers is the top element of $\mathsf{D}(\mathbb{G}^c)$.

At the other end of the determination ordering, we see that

Proposition 9.25.

1 For every specific number m: $m \preceq \alpha$ for every arbitrary number α (including the specific numbers).

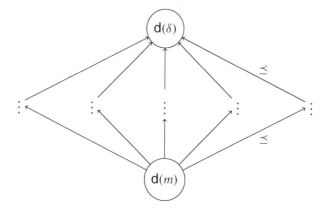

Figure 9.1 $D(\mathbb{G}^c)$.

2 For every specific number m and for every arbitrary number α: if $\alpha \preceq m$, then α is a specific number.

This of course means that the degree $d(m)$ of the specific natural numbers is the bottom element of $D(\mathbb{G}^c)$. In this way we obtain the picture in Figure 9.1.

9.4.2 Arbitrariness and randomness. It is doubtful that there is a unique clear and distinct property of randomness as applied to numbers that is singled out by our pre-theoretic intuitions of what it *means* for a number to be random. Nonetheless, there is a long history of attempts to distinguish between random and non-random numbers.

Some approaches regard randomness as a property of the way in which they are *produced*. For instance, one may call a real number random if its decimal expansion e has been generated by a *chance physical process* such as the repeated throwing of a die, or by *freely choosing* the successive digits of e. Some such idea is behind the intuitionistic theory of *choice sequences* (Troelstra 1977).

Other approaches focus not on the method of generation but on properties of its *outputs*, i.e. on properties of the sequences of digits that represent numbers. Approaches based on Kolmogorov complexity, as pioneered by Solomonoff (1960), are of this nature. One may then (roughly) call a natural number n random if the shortest computer programme that generates it in decimal notation contains more than n symbols. This approach can be extended to real numbers: an infinite string may be said to be random if the programme-size complexity of an initial segment of length n does not drop arbitrarily far below n (Chaitin 1975).

Let us now attempt to apply a concept of randomness to arbitrary natural numbers. Moreover, let us concentrate on the structure $\mathbb{G}^c$, or even drop the connection between states and ω-sequences altogether and focus on the structure $\mathbf{N}$ of Section 6.4.

Then we may wonder whether the ordering relation $\preceq$ in $\mathsf{D}(\mathbb{G}^c)$ tracks degree of randomness. If that were the case, then the diagonal numbers would be classified as being maximally random. As we shall see, however, it is not clear that this would be a desired outcome.

Let us reflect on what it could mean for one arbitrary natural number to be more random than another. One proposal latches on to the idea that for an arbitrary natural number to be random is for it to be *unconstrained*. This means first of all that a random number should be able to take any value, i.e. it should be *fully arbitrary* (Definition 4.2). But in a generic natural number structure, being unconstrained also has a *contextual* dimension. All other things being equal, an arbitrary number α is more random than an arbitrary number β if α can take some value $m \in \mathbb{N}$ in more circumstances – read: more *states* – than β.

This idea can be made precise in the following manner. We start with a definition:

Definition 9.26. The (modal) *profile* p_α of an arbitrary number α is the function such that for every $m \in \mathbb{N}$, $p_\alpha(m)$ is equal to the number of states in which α takes the value m.

Then it seems reasonable to require for an arbitrary number α to be said to be less random than or equally random as an arbitrary number β that for every $m \in \mathbb{N}$: $p_\alpha(m) \leq p_\beta(m)$. Consider therefore the relation

Definition 9.27. $\alpha \preceq_0^r \beta \equiv$ for every $m \in \mathbb{N} : p_\alpha(m) \leq p_\beta(m)$.

The relation $\preceq_0^r$ is reflexive and transitive, but not symmetrical. So the relation

$$x \preceq_0^r y \wedge y \preceq_0^r x$$

is an *equivalence relation* that partitions the arbitrary numbers α in equivalence classes $d_0^r(\alpha)$ that are partially ordered by $\preceq_0^r$. Call the resulting structure $\langle \mathsf{D}_0^r, \preceq_0^r \rangle$.

Unfortunately, the degree structure D_0^r is too fine-grained. It seems that our notion of randomness should be *permutation-invariant* in the following sense. Suppose that there exists a permutation $\pi : \mathbb{N} \to \mathbb{N}$ such that for all $m \in \mathbb{N}$, α takes the value m if and only if β takes the value $\pi(m)$. Let us abbreviate this as: $\beta = \pi(\alpha)$. In such circumstances, it seems intuitively compelling to say that α and β are equally random.

But the relation

$$\pi(x, y) \equiv \exists \text{ permutation } \pi : y = \pi(x)$$

is an equivalence relation on the arbitrary numbers. So we 'mod out' and consider the quotient structure

$$\langle \mathsf{D}^r, \preceq^r \rangle \equiv \langle \mathsf{D}_0^r/\pi, \preceq_0^r /\pi \rangle.$$

The result is a notion of degrees of randomness for arbitrary numbers that is neither concerned with the way they are generated (unlike the notions of randomness in intuitionistic mathematics), nor with the way they are expressed in notation systems (unlike the notions of randomness of algorithmic information theory).

It is easy to see that

Proposition 9.28.

1 The ω-generic numbers constitute the top element of the ordering $\langle \mathsf{D}^r, \preceq^r \rangle$;
2 The specific numbers constitute the bottom elements of the ordering $\langle \mathsf{D}^r, \preceq^r \rangle$.

In the context of the computable generic ω-sequence, this means that the constant functions are 'maximally random' numbers.

Because of the *contextual* aspect of the motivating idea behind this notion of randomness, the diagonal (or 1-generic) numbers are considered by $\langle \mathsf{D}^r, \preceq^r \rangle$ to be less random than the ω-generic numbers.

9.4.3 Independence. If we find the notion of determination on one end of the spectrum, then we find a natural notion of *independence* on the other end of the spectrum. Väänänen and Grädel motivate an extension of the framework of Dependence Logic in the following manner (Väänänen and Grädel 2013, p. 400):

We shall … give the concept of independence a similar treatment as we gave above to the concept of dependence. Again we start from the strongest conceivable form of independence of variables x and y, a kind of total lack of connection between them … We can read this in many ways:

- *x and y are completely independent from each other;*
- *x and y occur totally freely;*

- *x and y give absolutely no information of each other;*
- *every conceivable pattern occurs for x and y.*

This concept, too, can be applied to arbitrary numbers:

Definition 9.29. Two arbitrary numbers α and β are *independent* of each other (denoted as $\alpha \perp \beta$) if the respective values that they can take are always compatible with each other.

This relation can again be expressed in the language $\mathcal{L}$:

Proposition 9.30. For any $\alpha, \beta \in A$: $\alpha \perp \beta$ according to G iff

$$\mathsf{G} \models \forall^{\mathsf{Sp}} m \forall^{\mathsf{Sp}} n ((\Diamond(x = m) \wedge \Diamond(y = n)) \rightarrow \Diamond(y = m \wedge y = n))[x/\alpha, y/\beta].$$

It is easy to see that $\perp$ is symmetric but not transitive. On all arbitrary numbers except the specific ones, independence is irreflexive.

Proposition 9.4.3. Any two different constant functions c_i and c_j are independent of each other.

9.5 Computational Complexity

In this section we consider languages such as $\mathcal{L}$ and $\mathcal{L} \backslash \{\mathsf{Sp}\}$ and investigate how computationally complex the set of *generically true* sentences of such a language is. The reason why this question is interesting is that the complexity of truth sets is a standard measure of the expressive power of interpreted languages.

Let P and Q be subsets of $\mathbb{N}$. Recall from recursion theory that

Definition 9.31. P is *m-reducible* to Q, written $P \leq_m Q$, iff there exists a computable $f : \mathbb{N} \rightarrow \mathbb{N}$ such that for every $n \in \mathbb{N}$,

$$n \in P \iff f(n) \in Q.$$

Also, P and Q are called *m-equivalent*, written $P \equiv_m Q$, iff they are *m*-reducible to each other. The notion of *m*-reducibility plays a major role in studying complexity of decision problems and expressive power of formal languages in the foundations of mathematics.

The language of Peano Arithmetic was described earlier (Section 9.2.2). Let me therefore here restrict myself to reminding the reader of the definition of the language of second-order arithmetic. Without loss of generality, we can restrict our attention to *monadic* second-order arithmetic (since first-order arithmetic allows us to code elements of $\mathbb{N}^\ell$ as elements of $\mathbb{N}$). This language $\mathcal{L}_2$ includes two sorts of variables, namely

- *individual variables x, y, z, . . .* (intended to range over natural numbers);
- *set variables X, Y, Z, . . .* (intended to range over sets of natural numbers).

Accordingly, one must distinguish between *individual* and *set quantifiers,* viz.

$$\exists x, \exists y, \exists z, \ldots \quad \text{and} \quad \exists X, \exists Y, \exists Z, \ldots$$

The $\mathcal{L}_2$-*formulas* – or *monadic second-order σ-formulas* – are built up from the first-order σ_1-formulas and the expressions of the form $t \in X$, where t is a σ_1-term and X is a set variable, by means of the connective symbols and the quantifiers in the usual way. As usual, we write $\neg \exists X \neg \Phi$ as shorthand for $\forall X \Phi$, and adopt other standard abbreviations.

Let $\mathfrak{N}_2$ be the standard (or *full*) interpretation of the language $\mathcal{L}_2$ of (monadic) second-order arithmetic, where the monadic second-order quantifiers range over *all* sets of natural numbers. The set of first-order arithmetical truths and the set of second-order arithmetical truths are defined, as usual, as

Definition 9.32. $\mathrm{Tr}_1(\mathbb{N}) := \{\varphi \in \mathcal{L}_1 \mid \mathfrak{N} \models \varphi\}$.

Definition 9.33. $\mathrm{Tr}_2(\mathbb{N}) := \{\varphi \in \mathcal{L}_2 \mid \mathfrak{N}_2 \models \varphi\}$.

Then it can be shown that (Horsten and Speranski, 2018, Theorem 5.5):

Theorem 9.5.1.

$$Tr_{\mathcal{L}\backslash\{Sp\}}(\mathsf{G}) \equiv_m Tr_1(\mathbb{N}).$$

Theorem 9.5.2.

$$Tr_{\mathcal{L}}(\mathsf{G}) \equiv_m Tr_2(\mathbb{N}).$$

Recall that in these statements, G can be taken to be the computable generic ω-sequence or the full generic ω-sequence.

Theorem 9.5.1 says that the expressive power of $\mathcal{L}\backslash\{Sp\}$ is no larger than that of the language of first-order arithmetic: the intensional operator $\Box$ taken by itself does not add expressive power. So Kripke was right when he asserted (see Section 10.2) that a predicate to express a form of *rigidity* plays an essential role in increasing expressive power: as we have seen in Section 9.3, the predicate Sp expresses *structural rigidity*. Without the predicate Sp, interpreted as holding of the individual numbers, the language has *much* less expressive power than when the interpreted predicate Sp is included.

Let us now turn to Theorem 9.5.2. We have seen that in the computable generic ω-sequence $\mathbf{N}_C$ we find 2^ω arbitrary numbers (Section 9.2.1), and we will see in the next section that there is a precise sense in which arbitrary numbers can play the role of *sets of natural numbers*, or, equivalently, of *real numbers*. In view of this, it is not surprising that the computable generic ω-truths form a collection that is exactly as complicated as true second-order arithmetic. But there are 2^{2^ω} arbitrary numbers in the full generic ω-sequence $\mathbf{N}$ (see Section 6.3), i.e. as many as there are *sets* of real numbers. So it is less immediate that the truth set of the full generic ω-sequence is also exactly as complex as true second-order arithmetic rather than, for instance, as true third-order arithmetic or a substantial fragment thereof. But it turns out that it is. This provides some indication, by the way, that the full generic ω-sequence contains structure that, seen from a certain level of mathematical abstraction, does no work, and can therefore be considered redundant. In the next section, I provide additional reasons for preferring the computable generic ω-sequence over the full generic ω-sequence.

9.6 Axiomatising Second-Order Arithmetic

We have seen that the language $\mathcal{L}$ has much expressive power. But expressing facts about arbitrary numbers is one thing; *proving* facts about them is another. In this section I turn to the problem of how to axiomatise second-order number theory in $\mathcal{L}$.

9.6.1 Basic arithmetic. Recall that the symbols $0, 1, +, \times$ belong to the vocabulary of $\mathcal{L}$. Let us now see how the basic principles of arithmetic can be stated in $\mathcal{L}$.

We have seen in Section 9.3.1 that 0 and 1 can be explicitly defined in $\mathcal{L}$. These definitions can be expressed as axioms:

Axiom 9.34. $\forall x(x = 0 \leftrightarrow x + x = x)$

Axiom 9.35. $\forall x(x = 1 \leftrightarrow x \times x = x)$

The number 0 is the smallest number, and there are no loops in the numbers. These basic properties of the ordering of the natural numbers can be expressed as follows:

Axiom 9.36. $\forall x \forall y[x + y = 0 \rightarrow (x = 0 \wedge y = 0)]$

Axiom 9.37. $\forall x \forall y(x + 1 = y + 1 \rightarrow x = y)$

To conclude, there are the recursive axioms for $+$ and $\times$, which are postulated to hold for all arbitrary numbers:

Axiom 9.38. $\forall x(x + 0 = x)$

Axiom 9.39. $\forall x \forall y[x + (y + 1) = (x + y) + 1]$

Axiom 9.40. $\forall x(x \times 0 = 0)$

Axiom 9.41. $\forall x \forall y[x \times (y + 1) = (x \times y) + x]$

9.6.2 Comprehension and mathematical induction. We have seen that the first-order truth set of **G** has the complexity of second-order number theory. But the analogy between **G** and $\mathfrak{N}_2$ is even closer than that.

Arbitrary numbers can play the role of *sets* of specific natural numbers in a straightforward way. An arbitrary number α can be said to *code* the set $A \subseteq \mathbb{N}$ consisting of all and only the specific numbers $n \in \mathbb{N}$ such that α takes the value n in some state π.

This can of course be expressed in $\mathcal{L}$. We can *simulate* the elementhood relation in $\mathcal{L}$ as follows:

Definition 9.42. $x \in y =: \mathsf{Sp}(x) \wedge \Diamond(x = y)$.

This means that *specific* numbers play a double role as specific numbers and as singletons. Formally, for every $n \in \mathbb{N}$, we have $\mathbf{G} \models \mathbf{n} \in \mathbf{n}$. Note also that no arbitrary number plays the role of the empty set on this conception of the elementhood interpretation. Nonetheless, we can nevertheless simulate in $\mathcal{L}$ the role that the empty set plays in second order number theory: see Horsten and Speranski (2018, Section 5.1). Moreover, we have seen that if we would, as we have briefly entertained in Section 3.8.2, work with *partially defined* arbitrary natural numbers, then we would have a very natural candidate for representing the empty set in the form of the arbitrary natural number that is everywhere undefined.

Bearing Definition 9.42 in mind, we can express in $\mathcal{L}$ that every predicate determines a set:

Axiom 9.43 (Comprehension).

$$\exists y \forall^{\mathsf{Sp}} x[\Diamond(y = x) \leftrightarrow \Phi(x)],$$

where $\Phi \in \mathcal{L}$ does not contain y free.

This comprehension principle is easily seen to be true in **G**.

In this way, we reason about sets of numbers (or, equivalently, real numbers) in $\mathcal{L}$ in a natural way.

Let $x < y$ abbreviate $\exists^{Sp}z(x + z = y)$. The principle of second-order induction for specific natural numbers can then be expressed in $\mathcal{L}$ as follows:

Axiom 9.44.

$$\forall x : \{[\forall^{Sp}z(\forall^{Sp}y(y < z \rightarrow y \in x) \rightarrow z \in x)] \rightarrow \forall^{Sp}z(z \in x)\}$$

9.6.3 Second-order number theory in $\mathcal{L}$. We now have the ingredients for expressing the usual axioms of second-order Peano Arithmetic (PA^2) in $\mathcal{L}$. The resulting theory is denoted as $PA_{\mathcal{L}}^2$. It is defined as containing the following:

1 the logical axioms and rules 9.2–9.10 of Section 9.2.4;
2 the axioms 9.12–9.14 of Section 9.3 governing Sp;
3 the arithmetical axioms 9.34–9.41 of Section 9.6.1;
4 the comprehension axiom 9.43 of Section 9.6.2;
5 the induction axiom 9.44 of Section 9.6.2.

In other words, second-order number theory can be formalised in a natural way in $\mathcal{L}$. Clearly subsystems of second-order arithmetic can be expressed in the usual ways by imposing restrictions on comprehension axiom 9.43.

9.7 Well-Ordering

The theory $PA_{\mathcal{L}}^2$ is a standard second-order theory of specific natural numbers. But it is not a satisfactory (second order) theory of *arbitrary numbers*. In order to obtain a natural formal theory of arbitrary natural numbers in $\mathcal{L}$, some of the axioms need to be generalised.

In this section I want to concentrate on a particular difficulty in doing this. The standard way of proving facts about the natural numbers is by *mathematical induction*. But the arbitrary numbers are not well-ordered in any obvious way. So it is a non-trivial question whether a principle of induction governing all arbitrary numbers can be expressed in $\mathcal{L}$. I will argue that in this respect, the computable generic ω-sequence $\mathbb{G}^c$ is (again) more tractable than the full generic ω-sequence $\mathbb{G}$.

9.7.1 Well-ordering and transfinite mathematical induction. The principle of mathematical induction is the main engine for proving facts about the natural numbers. In set theory, transfinite induction is ubiquitously present in proofs about all ranks and therefore about all sets.

Induction presupposes well-ordering. But the arbitrary numbers are not obviously well-ordered. Indeed, Theorem 9.3.7 entails that even a total linear ordering of A, let alone a well-ordering of A, is undefinable in $\mathcal{L}$. This means that there is no straightforward way to formulate a principle of transfinite induction for proving facts about *all* arbitrary numbers. Of course we can express restricted principles of mathematical induction. Mathematical induction for the *specific* natural numbers, for instance, was formulated in Section 9.6.2. But this falls short of the mark.

At this point, a difference between $\mathbb{G}$ and $\mathbb{G}^c$ emerges. This difference is related to the diagonal numbers. I will now describe how *relative to a diagonal number*, a well-ordering of A can be defined. This helps us in the case of $\mathbb{G}^c$, which contains diagonal numbers even if no single one of them is definable in $\mathcal{L}$. But it does not help us in the case of the full generic system $\mathbb{G}$, which does not contain any diagonal numbers. I see this as another reason for preferring $\mathbb{G}^c$ over $\mathbb{G}$ for reasoning about arbitrary numbers.

Consider the *state space* Π^c of $\mathbb{G}^c$ and let δ be any diagonal number. The arbitrary number δ orders the elements of Π^c into an ω-sequence

$$\pi_0^\delta, \pi_1^\delta, \pi_2^\delta, \ldots,$$

where π_n^δ is the unique $\pi \in \Pi^c$ such that $\mathbb{G}^c \models_\pi (x = \mathbf{n})[x/\delta]$.

Now define a relation $<_\delta$ on the arbitrary numbers $\mathbb{A}^c$ of $\mathbb{G}^c$:

Definition 9.45. For all $f, g \in \mathbb{A}^c$:

$$f <_\delta g \Leftrightarrow:$$

for the least π_i^δ such that $f(\pi_i^\delta) \neq g(\pi_i^\delta)$, we have $f(\pi_i^\delta) < g(\pi_i^\delta)$.

It is clear that $<_\delta$ is a total strict linear ordering on Π^c. Moreover, we have:

Lemma 9.46. $<_\delta$ is a well-ordering of $\mathbb{A}^c$.

Proof Consider any $\mathcal{F} =: \{f_i : i \in I\} \subseteq \mathbb{A}^c$. Inductively define
(1) $\mathcal{F}_0 =: \mathcal{F}$;
(2) $\mathcal{F}_{n+1} =: \{f \in \mathcal{F}_n : \neg\exists f' \in \mathcal{F}_n \text{ such that } f'(\pi_n^\delta) < f(\pi_n^\delta)\}$,

and then set

$$\mathcal{F}_\omega =: \bigcap_n \{\mathcal{F}_n : n \in \mathbb{N}\}.$$

Then clearly $\mathcal{F}_\omega$ is a singleton $\{g\}$ with $g <_\delta f$ for all $f \in \mathcal{F}$ such that $f \neq g$. $\qquad\square$

Thus with every diagonal number $\delta \in \mathbb{A}^c$ a well-ordering of $\mathbb{A}^c$ is associated. In Section 9.3.2 we formulated a predicate $\mathsf{D}(x) \in \mathcal{L}$ that defines the class of diagonal numbers of $\mathbb{A}^c$. Moreover, it is clear that, relative to δ, we can define $<_\delta$ in $\mathcal{L}$. This allows us to define a principle of transfinite induction in $\mathcal{L}$.

Definition 9.47. A formula ϕ is *progressive* with respect to $<_\delta$ if and only if

$$\forall y : (\forall z <_\delta y : \phi(z)) \rightarrow \phi(y).$$

We denote this relation as $\mathsf{Prog}(\phi, <_\delta)$.

Similarly, in the light of comprehension axiom 9.43, we can express what it means for a *set* u of numbers to be progressive:

Definition 9.48. A set u is *progressive* with respect to $<_\delta$ if and only if

$$\forall y : (\forall z <_\delta y : z \in u) \rightarrow y \in u.$$

We denote this relation as $\mathsf{Prog}(u, <_\delta)$.

Then we can express in $\mathcal{L}$ a strong principle of *transfinite induction* for arbitrary numbers of $\mathbb{G}^c$:

Axiom 9.49 (Transfinite Induction).

$$\forall u[\exists x(\mathsf{D}(x) \wedge \mathsf{Prog}(u, <_x)) \rightarrow \forall y : y \in u].$$

This axiom is true in $\mathbb{G}^c$, and it covers all arbitrary natural numbers. Moreover, it entails the second-order induction for Peano Arithmetic, i.e. second-order induction for *specific* numbers. A fortiori, given comprehension axiom 9.43, it entails the induction axiom of first-order Peano Arithmetic.

10 | Probability and Random Variables

> As regards variables our results are as follows. Variable quantities may certainly
> be admitted, but do not belong to pure Analysis.
>
> *(Frege (1960, p. 111))*

Variables in mathematics and logic are symbols that can take different values; the old notion of variable as a kind of variable non-linguistic entity that was discussed in Section 3.2 has become obsolete.

Nonetheless, our discussion of fair dice in Section 6.2 suggested that there is an area where a version the old notion of variable is thriving: this is the area of probability theory, and the contemporary version of the old notion is that of a *random variable*.

In this chapter I discuss how a theory of random variables can be developed within the framework of arbitrary object theory. As before, the discussion is example-orientated. Moreover, the natural numbers again play a special role: particular attention is given to the way in which the generic ω-sequence can be seen as a collection of random variables.

10.1 Random Variables

A random variable is a numerical way of expressing the outcome of a random process in such a way that probabilities can be assigned to the outcome of the process lying in a numerical range.

Formally, a random variable X is a function from an outcome space Ω to a measurable space E (which is often taken to be $\mathbb{R}$) (Blitzstein and Hwang 2015, p. 92). The outcome space Ω is associated with a probability function: Ω is typically seen as a component of a probability triple $(\Omega, \mathfrak{A}, \mathsf{Pr})$ that is a classical probability space, i.e. where $\mathfrak{A} \subseteq \mathcal{P}(\Omega)$ is a σ-algebra on Ω and Pr satisfies the Kolmogorov axioms for probability (including the axiom of countable additivity).

Let us look at a simple example of modelling the stochastic behaviour of the fair coin.

Example 10.1. The throwing of a fair coin can be described in terms of a random variable X. Take the outcome space Ω to be the set of states

{fair coin coming up heads, fair coin coming up tails}.

Then a probability triple $(\Omega, \mathfrak{A}, \mathsf{Pr})$ can easily be defined in the Laplacian manner (with $\mathfrak{A}$ being the full power set of Ω) such that

$$\mathsf{Pr}(X = 1) = \mathsf{Pr}(X = 0) = \frac{1}{2}.$$

The understanding is that the standard coin is not *actually* in the state of (having been thrown and) having landed heads or in the state of (having been thrown and) having landed tails. The idea is that random variables *can* adopt values, but do not in actual fact have a determinate value. So a peculiar kind of modality is implicitly present here.

Often we associate different random variables with one object or with a system of objects. For instance, consider the throwing of two independent fair coins c_1 and c_2. The outcome space is now the set of four states

$$\{\langle c_1^+, c_2^+ \rangle, \langle c_1^+, c_2^- \rangle, \langle c_1^-, c_2^+ \rangle, \langle c_1^-, c_2^- \rangle\},$$

where c_i^+ stands for 'coin c_i landing heads', and c_i^- stands for 'coin c_i landing tails'. Then a random variable X can be defined as in example 10.1, describing the behaviour of the fair coin c_1, and a random variable Y describing the behaviour of the fair coin c_2 can be defined in a similar manner. But we can also define a third random variable Z that maps $\langle c_1^+, c_2^+ \rangle$ to 1 and maps all the other states of Ω to 0. The random variables X and Y can be given a straightforward physical interpretation: they represent the outcome of a 'random experiment' of throwing a fair coin. The random variable Z does not have such a straightforward physical interpretation, but it counts as a random variable nonetheless. Again Ω can be taken to be the sample space of a Laplacian probability triple $(\Omega, \mathfrak{A}, \mathsf{Pr})$. Then we see that

$$\mathsf{Pr}(X = 1) = \mathsf{Pr}(X = 0) = \mathsf{Pr}(Y = 1) = \mathsf{Pr}(Y = 0) = \frac{1}{2}.$$

We also see that for all $p, q \in \{0, 1\}$,

$$\mathsf{Pr}(X = p) = \mathsf{Pr}(X = p \mid Y = q),$$

i.e. the random variables X and Y are *independent* of each other. Furthermore, we see that the random variables X and Z are not independent. This seems a good description of the stochastic behaviour of the system of two independent fair coins.

The 'Laplacian' character of these situations consists in the facts that all the atomic events (singleton events) are given equal probability, so that if we have only a finite number of states, the *ratio formula* can be used. But of course this need not be the case: we may be considering biased coins, for instance. This is dealt with by assigning different *weights* to atomic events. Also, it is well known that for countably infinite sample spaces, there are no uniform probability distributions, whereas for uncountable sample spaces, there are (Finetti 1974). So, for instance, according to classical probability theory, the concept of a fair die with a countable infinity many faces is incoherent, whereas the concept of a fair die with uncountably many faces is perfectly coherent.

10.2 A Connection with Individual Concepts?

Kripke (1992, p. 72) has suggested that random variables can be seen as individual concepts. In this section I discuss the connection between random variables, individual concepts, and arbitrary objects.

In the previous section we saw how random variables are officially defined as *functions* of a certain kind. This means that random variables are not variables in the contemporary mathematician's or logician's sense: random variables are not symbols.

Random variables were introduced in statistics and probability theory for a purpose.[1] A random variable (also called '*aleatory* variable', or '*stochastic* variable') is a variable whose value is subject to variations due to chance (i.e. randomness in a mathematical sense). Random variables describe varying *quantities*.

The notion of a random variable is related to the concept of a *scientific* variable. To my surprise, I found it not easy to find attempted clear descriptions of how the notion of variable is understood in science (as opposed to mathematics or logic). The best we seem to get is statements along the following lines (Rovelli 2018, p. 3):

Classical mechanics describes the world in terms of physical variables. Variables take values, and these values describe the events of nature. Physical systems are characterized by sets of variables and interact. In the interaction, systems affect

---

[1] The history of the term and of the concept it expresses is complicated and interesting (see Shafer 2018): I will not go into it here.

one another in a manner that depends on the value taken by their variables. Given knowledge of some of these values, we can, to some extent, predict more of them.

The same does Quantum Mechanics.

The idea seems to be that a scientific variable is an *attribute* (or set of attributes) of an object (or a person or a system) that can vary. So there is a close connection between the concept of a scientific variable on the one hand, and the concept of an arbitrary property (see Section 3.9) on the other hand. A variable in the scientific sense is only a *random* variable if this variation can be described in probabilistic terms. The quantity is supposed to be *measurable*, and the values of the quantity are seen as *states* that the quantity can be in.

Like a random variable, an individual concept is a *function*, the domain of which is a collection of states (called *state descriptions*). But unlike a random variable, the range of an individual concept does not have a measure associated with it.

Like the concept of a random variable, the concept of an individual concept was introduced with a purpose in mind. Carnap (1956) introduced the term 'individual concept' as a term of art: individual concepts are supposed to serve as *meanings* (or intensions) for expressions that purport to refer to *objects*.[2] This shows that the idea behind individual concepts is not quite the same as the idea behind random variables. It would be natural, for instance, to describe the ways in which *one* given object changes colour by means of a random variable. In sum, random variables are concerned with the variation of *quantities*; individual concepts are concerned with the variation of denoted objects.

Individual concepts are in this sense more like arbitrary objects. But individual concepts are not quite the same as arbitrary objects. Arbitrary objects are multipurpose entities (see Section 4.6): they are not meant to serve one specific purpose before all others. Moreover, the ontological status of arbitrary objects differs from that of individual concepts. There is no agreement about whether meanings should be regarded as objects. But at least according to arbitrary object theory, arbitrary objects *are* supposed to be objects.

Actually, to say that individual concepts only serve as meanings for expressions is an oversimplification. In recent times we find suggestions that individual concepts can also be put to metaphysical use. Suppose

---

[2] I am ignoring here complications arising from non-denoting singular terms.

you believe in a four-dimensional theory of concrete objects (see e.g. Sider 2001). Then ordinary objects may be modelled as functions from moments in time to objects-at-a-time (time slices). In other words, on this metaphysical view ordinary objects can be modelled as individual concepts. Garson (2006, Section 13.8) calls this the *objectual* interpretation (as opposed to the conceptual interpretation) of individual concepts.

In sum, even though they are clearly related, random variables, individual concepts, and arbitrary objects are all different entities. Nonetheless, it may be fruitful to try to apply properties of one of them to an other of them. What I want to investigate is whether a concept of probability (which is essentially associated with random variables) can be applied to arbitrary objects. Unlike random variables, arbitrary objects do not 'come with' an associated probability function. Let us see whether there are natural ways to equip them with one.

10.3 Uniformity

As a simple test case, let us return to the example of the arbitrary hairdressers of Section 3.7.

One way to go would be simply to construct a probability measure $\langle K, \mathcal{P}(K), \mathsf{Pr} \rangle$ on the class K (with cardinality p) of specific hairdressers. Then if B is the class of hairdressers who have purple hair, $\mathsf{Pr}(B)$ can be taken to give the probability that a fully arbitrary hairdresser has purple hair.

This is not exactly what we want. We want to get a probability measure for every arbitrary hairdresser. But even if we restrict ourselves to *fully* arbitrary hairdressers – recall: there are $p!$ of them – we cannot obtain a different probability measure for each of those by conditionalisation of Pr to a subset of K. The problem is that we cannot capture how the different random variables are probabilistically correlated with each other.

I take this to be just another way of saying that use of random variables cannot be reduced in a straightforward way to talk about probability distributions only. In an attempt to exploit the analogy between arbitrary objects and random variables, I instead take the sample space of Pr to be the collection Ω of states in which an arbitrary hairdresser can be. Since Ω is finite, we can set the σ-algebra $\mathfrak{A}$ equal to $\mathcal{P}(\Omega)$.

Now there is only one right way to fix Pr: it is the *uniform distribution* over Ω, which is given by the ratio formula. The reason is that a fully arbitrary hairdresser σ is as likely one particular hairdresser m as another particular hairdresser n; the notion of *weight* (of states) does not figure in the theory of arbitrary hairdressers. So we have

$$\mathrm{Pr}(\sigma = m) = \mathrm{Pr}(\sigma = n) = \frac{1}{k}.$$

This is an important point of difference with random variables. A biased coin that has a probability of $\frac{1}{\pi}$ of landing heads can be described perfectly as a random variable. But there is no analogue of the biased coin among the arbitrary hairdressers.

Now we can capture the manner in which arbitrary hairdressers are probabilistically correlated. As in the case of random variables, we say that two arbitrary hairdressers σ_1 and σ_2 are *probabilistically independent* if, for all $a, b \in K$,

$$\mathrm{Pr}(\sigma_1 = a \mid \sigma_2 = b) = \mathrm{Pr}(\sigma_1 = a);$$

otherwise, σ_1, σ_2 are said to be *dependent* of each other. Then we find, as expected, that two fully arbitrary hairdressers that in each state differ from each other, nonetheless depend on each other. And we also see how we can find arbitrary hairdressers that are probabilistically independent of each other.

Example 10.2. Suppose that $K = \{k_1, k_2, k_3, k_4\}$, and that therefore $\Omega = \{s_1, s_2, s_3, s_4\}$. Consider the arbitrary hairdresser σ_1 such that $\sigma_1(s_1) = \sigma_1(s_2) = k_1$, and $\sigma_1(s_3) = \sigma_1(s_4) = k_2$, and the arbitrary hairdresser σ_2 such that $\sigma_2(s_1) = \sigma_2(s_4) = k_1$, and $\sigma_2(s_2) = \sigma_2(s_3) = k_2$. Then σ_1 and σ_2 are probabilistically independent of each other.

This indicates that the question of the *number* of pairs of independent arbitrary hairdressers relates to the divisibility of $|K|$. If $|K|$ is a prime number, for instance, then there are no such pairs.

For arbitrary objects with infinite value ranges, and therefore infinite associated state spaces, the requirement of uniformity presents a problem. As mentioned above, there are no uniform (classical) probability distributions on countably infinite sample spaces. Let us put questions about the *structure* of the natural numbers aside, and focus on the 'naive' way of modelling arbitrary numbers that was described in Section 4.1. Then $|\Omega| = \omega$, and then there is no uniform classical probability distribution on Ω.

In response to this, we could return to the full generic ω-sequence that was described in Section 6.3. Then $|\Omega| = 2^{\omega}$. There are uniform classical probability distributions on uncountable sample spaces: they assign probability 0 to each individual state in Ω. But this does not really solve out problem. It seems natural to demand of a truly arbitrary number σ that

$$\mathsf{Pr}(m) = \mathsf{Pr}(n)$$

for all $m, n \in \mathbb{N}$. But countable additivity again prevents this as a possibility.[3]

10.4 Non-Archimedean Probability Functions

Can we do better? Perhaps. I will now argue that we can if we instead adopt a non-Archimedean probability theory, some of our desiderata can be fulfilled.

In mathematics today, the term 'probability' has become virtually *synonymous* with 'function that satisfies the Kolmogorov axioms (including σ-additivity)'. If you see matters in this way, then you will be loath to dignify the functions constructed in the remainder of this chapter with the term 'probability'. Nonetheless, you may ask the question whether a *quantitative theory of possibility* can be constructed, by means of which degrees of possibility of properties can quantitatively be compared. This is the question that I will pursue. So, if you prefer, you can interpret the account that is developed in this chapter as a quantitative theory of afthairetic possibility. In that case you should mentally replace all occurrences of '(non-standard) probability function' by 'quantitative possibility function'.

I will focus on one particular non-Archimedean probability theory that is called NAP ('Non-Archimedean Probability'),[4] which I adapt to the present setting. In NAP, the unconditional probability of an event is defined in terms of the conditional probability of an event. Loosely speaking, the probability $\mathsf{Pr}(A)$ of event A is regarded as some kind of *limit* of the conditional probability $\mathsf{Pr}(A \mid \lambda)$, for the finite set λ 'tending toward infinity'. In other words, $\mathsf{Pr}(A)$ is conceived of the limit of the relative frequency of A's on finite snapshots of the sample space. As in previous chapters, I will focus on the case of the natural numbers.

---

[3] Some or all sets of states according to which $\sigma = m$ for some $m \in \mathbb{N}$ may not be in the σ-algebra $\mathfrak{A}$, but that also seems less than satisfying.

[4] This section relies heavily on the way in which this theory is described in Brickhill and Horsten (2018). The theory NAP was introduced in Benci et al. (2013); for a philosophical defence of this theory, see Benci et al. (2018).

10.4.1 Background. We start from the pre-structuralist treatment of arbitrary natural numbers of Section 4.1. This means that we take $\mathbb{N}$ to be both our state space and our outcome space. This gives rise to a collection

$$\mathsf{A}_n \equiv \{f : \mathbb{N} \to \mathbb{N}\}$$

of arbitrary natural numbers. Moreover, let δ denote some fixed diagonal number.

The arbitrary natural numbers are now to be treated as *random variables*: we want to associate a notion of probability with them. Our goal is to give precise meaning to conditional probability statements of the form

$$\mathsf{Pr}(\sigma \in A \mid \tau \in B),$$

where $\sigma, \tau \in \mathsf{A}_n$ and $A, B \subseteq \mathbb{N}$. It turns out that our way of making sense of statements of the form $\mathsf{Pr}(\sigma \in A \mid \tau \in B)$ can then in an obvious way be extended to make sense of more complicated statements such as

$$\mathsf{Pr}(R(\vec{\sigma}) \mid R'(\vec{\tau})),$$

where $\vec{\sigma} \in \mathsf{A}_n^k$ for some $k \in \mathbb{N}$, and $\vec{\tau} \in \mathsf{A}_n^l$ for some $l \in \mathbb{N}$.

Since Pr will take its values in a field, we can define $\mathsf{Pr}(\sigma \in A \mid \tau \in B)$ in the usual way as

$$\frac{\mathsf{Pr}(\sigma \in A \cap B)}{\mathsf{Pr}(\tau \in B)}.$$

Our fundamental problem therefore reduces to giving meaning to *unconditional* probability statements of the form $\mathsf{Pr}(\sigma \in A)$.

In accordance with Section 10.3, we insist on *uniformity*: we want every state to be equiprobable. Since δ takes every value exactly once, this requirement can be expressed as:

$$\mathsf{Pr}(\delta = m) = \mathsf{Pr}(\delta = n) \quad \text{for every } m, n \in \mathbb{N}.$$

10.4.2 The basic construction. I will describe how such a probability function can be built from an antecedently given fine free ultrafilter $\mathcal{U}$ on the collection $[\mathbb{N}]^{<\omega}$ of finite subsets of $\mathbb{N}$. So it is useful to label a probability function of the kind that we will create with the *ultrafilter* from which it is generated. I will therefore from now on denote probability functions of the kind that we are interested in as $\mathsf{Pr}_{\mathcal{U}}$.

I assume the notion of a free ultrafilter is to be known to the reader; I here merely pause to recall what it means to be a *fine* ultrafilter:

Definition 10.3. A *fine* ultrafilter on $[S]^{<\omega}$ is an ultrafilter $\mathcal{U}$ such that for every $x \in S$:

$$\{T \in [S]^{<\omega} : x \in T\} \in \mathcal{U}.$$

So let us start with a fine ultrafilter $\mathcal{U}$ on $[\mathbb{N}]^{<\omega}$, where $\mathbb{N}$ is regarded as the state space. Fine-ness is easily seen to imply free-ness, so $\mathcal{U}$ will then be a free ultrafilter as well. This fine ultrafilter $\mathcal{U}$ defines a non-archimedean field $\mathcal{F}_{\mathcal{U}}$ in the following way.

For any two functions $f, g : [\mathbb{N}]^{<\omega} \to \mathbb{Q}$ we define the following:

Definition 10.4.

$$f \approx_{\mathcal{U}} g \equiv \{T \in [\mathbb{N}]^{<\omega} : f(T) = g(T)\} \in \mathcal{U}.$$

In words: two functions are identified if they coincide on ultrafilter-many states.

The relation $\approx_{\mathcal{U}}$ is an equivalence relation, so we can take equivalence classes $[\ldots]_{\mathcal{U}}$ for which we then have

$$[f]_{\mathcal{U}} = [g]_{\mathcal{U}} \Leftrightarrow f \approx_{\mathcal{U}} g.$$

Moreover, it is a routine exercise to verify that the $[f]_{\mathcal{U}}$'s form a hyper-rational field $\mathcal{F}_{\mathcal{U}}$ when the usual field operations are defined pointwise (Goldblatt 1998, Chapter 3).

Now suppose $A \subseteq \mathbb{N}$ and $\sigma \in \mathbf{A}_n$. Then we define the function $f_{\sigma \in A} : [\mathbb{N}]^{<\omega} \to \mathbb{Q}$ as follows:

Definition 10.5. For every $T \in [\mathbb{N}]^{<\omega}$:

$$f_{\sigma \in A}(T) \equiv \frac{|\{s \in T : \sigma(s) \in A\}|}{|T|}.$$

In words: for every finite set of states T, $f_{\sigma \in A}(T)$ is the finite ratio between the number of states s in T for which $\sigma(s) \in A$ and the number of states in T. In this sense, $f_{\sigma \in A}(T)$ is the probability of $\sigma \in A$ *on a finite snapshot of states*.

Now we are ready to define the probability of $\sigma \in A$, relative to a fine (and therefore free) ultrafilter $\mathcal{U}$ on $[\mathbb{N}]^{<\omega}$:

Definition 10.6.

$$\mathrm{Pr}_{\mathcal{U}}(\sigma \in A) \equiv [f_{\sigma \in A}]_{\mathcal{U}}.$$

This means that $\text{Pr}_{\mathcal{U}}(\sigma \in A)$ is conceived of as the 'limit' (in a generalised sense) of the function $f_{\sigma \in A}$ in the hyper-rational field $\mathcal{F}_{\mathcal{U}}$.

10.4.3 Basic properties. Let us now review some basic pleasing properties that are shared by all probability functions that are constructed in this way. The (easy) proofs of the propositions concerning these properties that we will look at can be found in Benci et al. (2013).

Proposition 10.7. $\text{Pr}_{\mathcal{U}}$ is a finitely additive probability function.

Proposition 10.8. $\text{Pr}_{\mathcal{U}}$ is total.

Proposition 10.7 is of course a minimum condition that needs to be satisfied if we want to call $Pr_{\mathcal{U}}$ a probability function. Proposition 10.8 says that all subsets of $\mathbb{N}$ are measurable according to $\text{Pr}_{\mathcal{U}}$.

Next, we will see that the probability function $Pr_{\mathcal{U}}$ is fine-grained. For this, we first introduce some new terminology:

Definition 10.9. A probability function $Pr_{\mathcal{U}}$ is *Euclidean* if and only if for all $A, B \subseteq \mathbb{N}$:

$$A \subsetneq B \Rightarrow \text{Pr}_{\mathcal{U}}(\delta \in A) < \text{Pr}_{\mathcal{U}}(\delta \in B).$$

Then we see the following:

Proposition 10.10. $\text{Pr}_{\mathcal{U}}$ is Euclidean.

This shows that our probability measures $\text{Pr}_{\mathcal{U}}$ are fundamentally different from the notion of cardinality, which does not satisfy the Euclidean property.

Definition 10.11. A probability function $Pr_{\mathcal{U}}$ is *regular* if and only if $\forall A \in \mathcal{P}(\Omega) \setminus \{\emptyset\}$:

$$\text{Pr}(\delta \in A) > 0$$

Proposition 10.12. $Pr_{\mathcal{U}}$ is regular.

Proof   The argument that shows this makes crucial use of the *fine-ness* of $\mathcal{U}$. $\square$

Regularity is considered to be a desirably property because it means that the probability function distinguishes 'remote possibilities' from impossibilities.

In Section 10.3 it was argued that *uniformity* should be a property of probability functions associated with arbitrary objects seen as random variables, where uniformity was defined as follows:

Definition 10.13. A probability function $Pr_{\mathcal{U}}$ is *uniform* if and only if

$$\mathsf{Pr}(\delta = m) = \mathsf{Pr}(\delta = n) \quad \text{for every } m, n \in \mathbb{N}.$$

The probability functions that we have constructed meet this requirement:

Proposition 10.14. $Pr_{\mathcal{U}}$ is uniform.

Observe that this last proposition entails that $Pr_{\mathcal{U}}$ cannot be a countably additive probability function, for this would contradict proposition 10.10.

Countable additivity means that the probability of the union of a countable family of disjoint sets is the *infinite sum* of the probabilities of the elements of the family, where the notion of infinite sum is spelled out in terms of the classical notion of limit. In the present setting, the probability $Pr_{\mathcal{U}}$ of the union of *any* family of disjoint sets is also the infinite sum of the probabilities of the elements of the family (Benci et al. 2013, Section 3.4). But now the notion of infinite sum is spelled out in terms of the generalised notion of limit based on the ultrafilter $\mathcal{U}$. More precisely, the new notion of infinite sum is defined as follows. Suppose we are given a family $\{q_i : i \in \mathbb{N}\}$ of rational numbers. Then consider the function

$$f : [\mathbb{N}]^{<\omega} \to \mathbb{Q} : f(T) = \sum_{I \cap T} q_i.$$

This function can be seen as giving the value of the infinite sum on all *finite parts* ('snapshots') of the index set. So we identify the infinite sum of the family $\{q_i : i \in \mathbb{N}\}$ of rational numbers with the generalised limit of f according to the ultrafilter $\mathcal{U}$:

Definition 10.15.

$$\sum_{i \in I}^{*} q_i \equiv [f]_{\mathcal{U}}.$$

Using this notion of infinite sum, we can express the probability of a family of sets as the sum of the probabilities of the members of the family:

Proposition 10.16. If $A = \bigcup_{i \in I} A_i$, with $A_i \cap A_j = \emptyset$ for all $i, j \in I$, then:

$$\mathsf{Pr}_{\mathcal{U}}(\sigma \in A) = \sum_{i \in I}^{*} \mathsf{Pr}_{\mathcal{U}}(\sigma \in A_i).$$

In sum, $\mathsf{Pr}_{\mathcal{U}}$ has a natural infinite additivity property that is sometimes called *perfect additivity*.

All this sounds very satisfying. But things are not as good as they appear. The properties of an NAP function are known to be very sensitive to the choice of the ultrafilter on which it is based (Kremer 2014; Benci et al. 2018, Section 6.1).

We should not mind *some* sensitivity to choice of ultrafilter. After all, there is no reason to believe that there is only and only one correct probability function on A_n. But at the same time, we may want the sensitivity to choice of ultrafilter to be constrained in various ways. For instance, if we want to obtain, for δ being a diagonal number, and E is the set of even numbers and $\overline{E}$ is its complement, that

$$\mathsf{Pr}(\delta \in E) = \mathsf{Pr}(\delta \in \overline{E}),$$

then additional conditions have to be imposed on the class of admissible ultrafilters. It is not clear that *all* intuitively desirable properties for Pr can be guaranteed to hold by narrowing the class of ultrafilters on which probability functions can be based in a suitable way.

10.5 Set Probabilities

I will now extend the approach of the previous section to set theory: the aim is to apply the NAP conception of probability to arbitrary (pure) *sets*. In this section, I rely in parts on results from Brickhill and Horsten (2019).

10.5.1 The aim. Despite the exceptional position of set theory (see Section 5.7), let us start from the generic system of sets in the way that we would do for any other kind of object (see Section 3.7). This means that we take the set theoretic universe V not only as outcome space but also as state space, and have the functions $\sigma : V \to V$ as arbitrary sets, which will be treated as *random variables* over the set theoretic universe. The resulting collection of all arbitrary sets is denoted as A_s.

Our goal is again to give precise meaning to conditional probability statements of the form

$$\mathsf{Pr}(\sigma \in A \mid \tau \in B),$$

where $\sigma, \tau \in \mathsf{A}_s$ and $A, B \subseteq V$. But as before it will be sufficient for our purposes to give meaning to *unconditional* probability statements of the form $\mathsf{Pr}(\sigma \in A)$. So our fundamental problem amounts to giving meaning to expressions of the form $\mathsf{Pr}(\sigma \in A)$. As you will have come to expect by now, such probability measures will be determined by a choice of a fine ultrafilter on the collection $[V]^{<\omega}$ of finite subsets of the state space.

At this point you may be concerned about this involvement with proper classes.[5] But strongly inaccessible ranks are 'full' models of set theory. So we may trade the involvement with proper classes for a modest large cardinal assumption, and instead talk about elements of $V_{\kappa+1}$, where κ is a strongly inaccessible cardinal number. Indeed, for probability on *any* large set S, the general idea is the same as that for probability on V: the probability of A on S is determined by the set of probabilities of $\varphi(x)$ on 'small' subsets of S, where being entails being of smaller cardinality'. With this disclaimer in place, I will from now on continue to speak about the probability of an arbitrary set σ belonging to a proper class A, and I will sometimes even write '$|A| < |B|$' where A is a set and B is a proper class. Indeed, the attentive reader will have noticed that earlier in this section I already made use of this abuse of notation, when I wrote '$A, B \subseteq V$'.

10.5.2 The basic construction. We can assign probabilities to statements about arbitrary sets in much the same way as we did in the previous section for statements about arbitrary natural numbers.

The starting point is a fine ultrafilter $\mathcal{U}$ on $[V]^{<\omega}$. (As was explained in the previous subsection, if you are uncomfortable with an ultrafilter on $[V]^{<\omega}$, then take instead an ultrafilter on $[V_\kappa]^{<\omega}$, where κ is a strongly inaccessible cardinal.) This fine ultrafilter $\mathcal{U}$ defines a non-Archimedean field $\mathcal{F}_{\mathcal{U}}$ in the following way.

For any two functions $f, g : [V]^{<\omega} \to \mathbb{Q}$ we define the following:

Definition 10.17.

$$f \approx_{\mathcal{U}} g \equiv \{T \in [V]^{<\omega} : f(T) = g(T)\} \in \mathcal{U}.$$

In words: two functions are identified if they coincide on ultrafilter-many states.

The relation $\approx_{\mathcal{U}}$ is an equivalence relation, so we can take equivalence classes for which we then have

$$[f]_{\mathcal{U}} = [g]_{\mathcal{U}} \Leftrightarrow f \approx_{\mathcal{U}} g.$$

Moreover, it is again a routine exercise to verify that the $[f]_{\mathcal{U}}$'s form a hyper-rational field $\mathcal{F}_{\mathcal{U}}$.

---

[5] I am not, but I will not take a stance on the existence of proper classes here.

Now suppose $A \subseteq V$ and $\sigma \in \mathsf{A}_s$. Then we define the function $f_{\sigma \in A}$: $[V]^{<\omega} \to \mathbb{Q}$ as follows:

Definition 10.18. For every $T \in [V]^{<\omega}$:

$$f_{\sigma \in A}(T) \equiv \frac{|\{s \in T : \sigma(s) \in A\}|}{|T|}.$$

In words: for every *finite* set of states T, $f_{\sigma \in A}(T)$ is the ratio between the number of states s in T for which $\sigma(s) \in A$ and the number of states in T. In this sense, $f_{\sigma \in A}(T)$ is the probability of $\sigma \in A$ *on a finite snapshot of states*.

Now we are ready to define the probability of $\sigma \in A$, relative to a fine (and therefore free) ultrafilter $\mathcal{U}$ on $[V]^{<\omega}$:

Definition 10.19.

$$\mathsf{Pr}_{\mathcal{U}}(\sigma \in A) \equiv [f_{\sigma \in A}]_{\mathcal{U}}.$$

Thus we have constructed a probability function $\mathsf{Pr}_{\mathcal{U}}$ that takes its values in the hyper-rational field $\mathcal{F}_{\mathcal{U}}$.

10.5.3 Global properties. I will now give a brief overview of properties that probability measures $\mathsf{Pr}_{\mathcal{U}}$ share if $\mathcal{U}$ is a fine ultrafilter on $[V]^{<\omega}$.

The probability measures $\mathsf{Pr}_{\mathcal{U}}$ all have a number of pleasing properties. As in the case of non-Archimedean probability functions governing arbitrary natural numbers, it is a routine matter to verify that

Lemma 10.20.

1 $\mathsf{Pr}_{\mathcal{U}}$ is a finitely additive probability function;
2 $\mathsf{Pr}_{\mathcal{U}}$ is Euclidean;
3 $\mathsf{Pr}_{\mathcal{U}}$ is total;
4 $\mathsf{Pr}_{\mathcal{U}}$ is regular;
5 $\mathsf{Pr}_{\mathcal{U}}$ is uniform;
6 $\mathsf{Pr}_{\mathcal{U}}$ is perfectly additive.

But the Euclidean-ness of $\mathsf{Pr}_{\mathcal{U}}$ has implications for *symmetry principles*. As a rule of thumb, you can say that *symmetry principles fail*.[6]

Proposition 10.21. *$Pr_{\mathcal{U}}$ is not invariant under all permutations of V.*

---

[6] See Benci et al. (2007) and Benci et al. (2013, 2018).

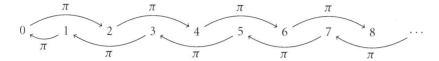

Figure 10.1 Permutation π.

Proof We concentrate on $\mathbb{N}$ as it is canonically represented in V (by means of the Zermelo ordinals, for instance). We define a permutation π of V as follows (see Figure 10.1):

- $\pi(x) = x$ for $x \in V \setminus \mathbb{N}$; Otherwise:
- $\pi(0) = 2$;
- $\pi(1) = 0$;
- $\pi(x) = x + 2$ for x even and > 0;
- $\pi(x) = x - 2$ for x odd and > 1.

Let $A \equiv \{0, 2, 4, \ldots\}$. Then $\pi(A) \subsetneq A$. Therefore, by the Euclidean principle, $\mathsf{Pr}_\mathcal{U}[\sigma \in \pi(A)] < \mathsf{Pr}_\mathcal{U}[\sigma \in A]$. $\qquad\square$

One popular global constraint on probability measures is *translation-invariance*. The Lebesgue measure has this property, and Banach limits seem to occupy a privileged position in the class of generalised limits at least in part because they are translation-invariant. In our context, translation-invariance does not make obvious sense. If you take a random class A, then it is not clear what '$A + \alpha$' (where α is a number) *means*. But a clear interpretation of 'adding an ordinal number' can of course be given if A is a collection of ordinals:

Definition 10.22. For A any collection of ordinals:

$$A \oplus \alpha \equiv \{\beta : \exists \gamma \in A \text{ such that } \beta = \gamma + \alpha\}.$$

Then for A to be translation-invariant means that for all ordinals α,

$$\mathsf{Pr}_\mathcal{U}(A) = \mathsf{Pr}_\mathcal{U}(A \oplus \alpha).$$

(Note that the operation $\oplus$ should not be confused with ordinal addition ($+$).)

However, even if we consider non-Archimedean measures (of the kind that we have been describing) on ordinals, translation-invariance conflicts with the Euclidean Property of our generalised probability functions. In particular, there is no NAP probability function $\mathsf{Pr}_\mathcal{U}$ on an infinite cardinal κ such that for all ordinals there is an ordinal α with $0 < \alpha < \kappa$ with

$$\mathrm{Pr}_{\mathcal{U}}(\delta \in \kappa) = \mathrm{Pr}_{\mathcal{U}}(\delta \in \kappa \oplus \alpha).$$

The reason is simple. We have $\kappa \oplus \alpha = \kappa \backslash \alpha \subsetneq \kappa$, so if we had $\mathrm{Pr}_{\mathcal{U}}(\delta \in \kappa) = \mathrm{Pr}_{\mathcal{U}}(\delta \in \kappa \oplus \alpha)$, then we would contradict the Euclidean principle.

Benci, Forti, and Di Nasso (2007, Section 1.3) explore a *restricted* notion of translation-invariance of NAP-like measures on ordinals. I do not pursue this theme further here, but only pause to note that there are other reasonable-looking principles that are hard to satisfy. In the context of their theory of *numerosities*, Benci, Forti, and Di Nasso consider a principle that in the present context would take the following form:

Axiom 10.23 (Difference Principle).

$$\forall A, B \in V : \mathrm{Pr}_{\mathcal{U}}(\delta \in A) < \mathrm{Pr}_{\mathcal{U}}(\delta \in B) \Rightarrow$$
$$\exists C \in V : \mathrm{Pr}_{\mathcal{U}}(\delta \in B) = \mathrm{Pr}_{\mathcal{U}}(\delta \in A) + \mathrm{Pr}_{\mathcal{U}}(\delta \in C).$$

On countable sample spaces, the difference principle can be made to hold by building Pr from a *selective* ultrafilter (Benci and Di Nasso 2003). But the existence of selective ultrafilters is independent of ZFC. As far as I know, it is an open whether the difference principle can be consistently made to hold for NAP probability functions on uncountable sample spaces.

10.5.4 Probability and cardinality. Cardinality is a measure of size. Our non-Archimedean probability functions $\mathrm{Pr}_{\mathcal{U}}$ purport at least to be *measures* of some kind. Theories of size in the spirit of the theory NAP have sometimes been proposed as rivals to Cantor's theory of cardinality, i.e. they have been presented as alternative measures of *size* of sets. This is done, for instance, in Benci et al. (2007).

I have no quibbles with Cantor's theory of cardinality. So I do not regard the theory NAP as an alternative measure of size of sets. Nonetheless, it is natural to wonder whether there are general principles that connect cardinality with our notion of non-Archimedean probability. This is what I will be looking at in the present (sub-)section.

Some principles connecting cardinality with non-Archimedean probability that come to mind, turn out to fail. Others can be made to hold. Where this is so, they are made to hold by imposing *extra restrictions* (beyond fine-ness) on the class of ultrafilters from which the probability functions are built. The proofs of the propositions to that effect that are discussed below, are given in Brickhill and Horsten (2019).

10.5.4.1 Hume's Principle for Probability

You might wonder whether the following probabilistic analogon of Hume's Principle for cardinality holds:

Axiom 10.24 (Hume's principle for probability). For all $A, B \in V$, and for all $\sigma \in A_s$:

$$|A| = |B| \Rightarrow \mathsf{Pr}_\mathcal{U}(\sigma \in A) = \mathsf{Pr}_\mathcal{U}(\sigma \in B).$$

But the probability functions $\mathsf{Pr}_\mathcal{U}$ that we have been considering cannot satisfy axiom 10.24, for it is in immediate conflict with the Euclidean Principle that they do satisfy (Lemma 10.20). However, this was only to be expected, and is not to be lamented. After all, we do not expect Kolmogorov probability (on infinite spaces) to satisfy any such principle.

10.5.4.2 Superregularity

The hyper-rational field $\mathcal{F}_\mathcal{U}$ in which the probability functions $\mathsf{Pr}_\mathcal{U}$ take their values contain *infinitesimal numbers* – this is what makes it non-Archimedean. So I will write $\mathsf{Pr}_\mathcal{U}(\sigma \in A) \approx 0$ if $\mathsf{Pr}_\mathcal{U}(\sigma \in A) < n^{-1}$ for each $n \in \mathbb{N}$. And I will write $\mathsf{Pr}_\mathcal{U}(\sigma \in A) \ll \mathsf{Pr}_\mathcal{U}(\tau \in B)$ if

$$\frac{\mathsf{Pr}_\mathcal{U}(\sigma \in A)}{\mathsf{Pr}_\mathcal{U}(\tau \in B)} \approx 0.$$

We have seen that $\mathsf{Pr}_\mathcal{U}$ cannot satisfy Hume's principle. But, at least at first sight, it seems that it would be reasonable to demand

$$|A| < |B| \Rightarrow \mathsf{Pr}_\mathcal{U}(\delta \in A) < \mathsf{Pr}_\mathcal{U}(\delta \in B).$$

Indeed, if, in addition, $|A| \geq \omega$, then we might expect

$$\mathsf{Pr}_\mathcal{U}(\sigma \in A) \ll \mathsf{Pr}_\mathcal{U}(\sigma \in B).$$

Also, this is expected to hold even if B is a proper class. The result is a size constraint which is a strengthening of the requirement of regularity (Definition 10.11):

Axiom 10.25 (Superregularity).

$$\omega \leq |A| < |B| \leq |V| \Rightarrow \mathsf{Pr}_\mathcal{U}(\delta \in A) \ll \mathsf{Pr}_\mathcal{U}(\delta \in B).$$

By a suitable restriction on admissible ultrafilters $\mathcal{U}$, superregularity can also be made to hold:

Lemma 10.26. There are fine ultrafilters $\mathcal{U}$ such that $\mathsf{Pr}_\mathcal{U}$ is superregular.

We have seen that Hume's Principle for probability (axiom 10.24) cannot hold for the notion of probability that we are investigating. But this leaves open the question whether the *converse* of Hume's Principle for probability can be made to hold. This is called *Cantor's Principle* in Benci et al. (2007), where the authors investigate it in the context of their Euclidean theory of size:

Axiom 10.27 (Cantor's Principle).

$$\mathsf{Pr}_{\mathcal{U}}(\delta \in A) = \mathsf{Pr}_{\mathcal{U}}(\delta \in B) \Rightarrow |A| = |B|.$$

Benci, Forti, and Di Nasso prove that 'Cantor's Principle' can be made to hold (Benci et al. 2007, Section 3.2). It is also clear that Cantor's Principle follows from superregularity (axiom 10.25).

10.5.4.3 The power set principle

It is well known that the question whether

$$\forall A, B \in V : |A| < |B| \Rightarrow |\mathcal{P}(A)| < |\mathcal{P}(B)|$$

is independent of the axioms of set theory. (Of course the principle is true if the generalised Continuum Hypothesis holds.)

Like the cardinality operator, our NAP probability operator is a measure. One might wonder what should follow from $\mathsf{Pr}_{\mathcal{U}}(\delta \in A) < \mathsf{Pr}_{\mathcal{U}}(\delta \in B)$. In particular, given that $\mathsf{Pr}_{\mathcal{U}}$ is intended to be a *fine-grained* quantitative possibility measure, perhaps probability should be expected to co-vary with the power set operation in some fairly direct manner. In other words, it is natural to ask if the following principle can be made to hold:

Axiom 10.28 (Power Set Condition).

$$\forall A, B \in V : \mathsf{Pr}_{\mathcal{U}}(\delta \in A) < \mathsf{Pr}_{\mathcal{U}}(\delta \in B) \Leftrightarrow \mathsf{Pr}_{\mathcal{U}}(\delta \in \mathcal{P}(A)) < \mathsf{Pr}_{\mathcal{U}}(\delta \in \mathcal{P}(B)).$$

Let us call this the *Power Set Condition*.

It turns out that axiom 10.28 can indeed be satisfied:

Theorem 10.29. There are ultrafilters $\mathcal{U}$ such that $\mathsf{Pr}_{\mathcal{U}}$ satisfies the power set condition.

10.5.5 Probabilities and ordinals. For $\alpha \geq \omega$, in each level $V_{\alpha+1} \setminus V_{\alpha}$ of the iterative hierarchy, one finds only *one* ordinal, but infinitely many sets that are not ordinals. This might lead one to believe that a probability function on V should satisfy

$$\mathsf{Pr}_{\mathcal{U}}(\delta \in On) \approx 0,$$

where *On* is the property of being an ordinal.

Just as it seems reasonable to require that the probability of choosing an even natural number from the set of natural numbers must be equal to or infinitesimally close to $\frac{1}{2}$, it seems reasonable to require that

$$\mathsf{Pr}_\mathcal{U}[\delta \in \text{Even} \mid \delta \in On] \approx \frac{1}{2},$$

where Even is the property of being an even ordinal, which is defined in the obvious way as being of the form $\lambda + n$ for λ a limit ordinal and n an even finite ordinal.

Moreover, between any two limit ordinals there are infinitely many successor ordinals, so one might expect

$$\mathsf{Pr}_\mathcal{U}[\delta \in \text{Lim} \mid \delta \in On] \approx 0,$$

where Lim is the property of being a limit ordinal.

Non-Archimedean probability functions can indeed be constructed that meet these expectations. Indeed, there are probability functions that meet these 'ordinal expectations' and in addition meet the size constraint of superregularity.

Theorem 10.30. There are superregular NAP probability functions $\mathsf{Pr}_\mathcal{U}$ such that

1 $\mathsf{Pr}_\mathcal{U}(\delta \in On) \approx 0$;
2 $\mathsf{Pr}_\mathcal{U}(\delta \in \text{Even} \mid \delta \in On) \approx \frac{1}{2}$;
3 $\mathsf{Pr}_\mathcal{U}(\delta \in \text{Lim} \mid \delta \in On) \approx 0$.

We have seen in Section 4.6 how Fine, in his 1985 book on arbitrary object theory, lists problems for future research. In this final and very brief chapter I follow Fine's example. I reflect on problems in arbitrary object theory that, in my opinion and as far as I can tell, merit philosophers' and logicians' attention.

The questions below constitutes no more than somewhat specific gestures at directions of possible future research. I may well be wrong about some of the questions below: some of them may be ill-conceived or unrewarding. But I am convinced that an enormous amount of philosophical and logical work remains to be done.

However, before turning to possible avenues for further research, I will briefly reflect on how what I have done in this monograph should be evaluated.

11.1 Success or Failure?

This book has been an exercise in naive metaphysics. Metaphysics can be done well, and it can be done badly. Early in this book (at the end of Section 2.4), I claimed that we *recognise* good metaphysics as good metaphysics when we see it. But more can be said. There are *criteria* for distinguishing good from bad metaphysics. These criteria are necessarily somewhat vague, but they are not completely uninformative.

Unlike in science, fit with precise empirical predictions and observations plays only a very limited role in the metaphysical methodology. Instead, it is a mark of good metaphysics that it reflects and gives insight into metaphysical appearance.

Accordance with metaphysical experience is by no means an empty condition. Consider the claim, made in Section 3.7, that for an arbitrary F, considered on its own, there are at least as many states as are necessary for there to be fully arbitrary F's. Or consider the claim that for any arbitrary F, call it a, there is another arbitrary F, call it b, such that, necessarily,

$a \neq b$. I submit that these conditions are mandated by how arbitrary F's metaphysically appear to us.

But faithfulness to direct metaphysical appearance does not carry us very far. Consider, for instance, the claim, also made in Section 3.7, that for an arbitrary F, considered on its own, there are no more states than are needed for there to be diagonal numbers. This can be seen as an application of Ockham's razor. It seems that this claim outstrips direct metaphysical experience.

So there is a role for *theorising* in metaphysics. And some of the methodological principles that are used in science, have a place in metaphysics as well. The goal of this theoretical component is to deepen our *understanding* of the metaphysical phenomena.

Pointed and searching metaphysical questions, such as those that were asked by Frege (see Section 4.2), are of key importance. They are *research problems* in something like Lakatos' (1970) sense of the word. If plausible answers to searching questions can be given, if our enquiry into pointed questions lead us to a richer metaphysical account that seems to offer deeper understanding, then that is a good sign. It means that we are making progress in our *metaphysical* research programme. In other words, Fregean questions can serve as *catalysts* in the development of arbitrary object theory.

If the answers that the metaphysician comes up only pose even more serious metaphysical questions, then that is a worrying sign. Let me illustrate, by means of a few examples, how this applies to Fine's theory of arbitrary objects and to the account that I have developed and advocated in this book. A specific modality that I have called afthairetic possibility plays a fundamental role in my theory of arbitrary objects. If answers to basic questions concerning this modality – about its nature, about its relations to other modalities, and so on – are not forthcoming or are fundamentally unsatisfactory, then that is a problem for my account. An irreducible notion of dependence plays a central role in Fine's theory of arbitrary objects. If basic questions concerning the nature of this relation and concerning its properties cannot be answered in at least a promising way, then that is not a good sign.

There is a perception that *applications* of arbitrary object theory are of great methodological importance: if arbitrary object theory can shed philosophical light on phenomena that are not *obviously* concerned with arbitrary objects, then that speaks in its favour (see Section 4.6). This is the metaphysical counterpart of the theoretical virtue of *fruitfulness* in scientific methodology (Kuhn 1977). But the naive metaphysician is interested in arbitrary objects in themselves, even in the absence of 'applications'. So paucity of philosophical applications is not a sufficient reason to reject arbitrary objects.

Along these lines we can evaluate individual metaphysical theories (such as Fine's and mine) and research programmes (such as arbitrary object theory as a whole). It is probably wise for me to leave it to others to debate the merits and demerits of my attempt to give a naive metaphysical account of arbitrary objects, arbitrary systems, and arbitrary numbers.

11.2 Towards a General Theory

Reflecting on the idea of variable numbers, Church (1962, p. 13) had this to say:

Objections to the idea that real numbers are to be divided into two sorts or classes, 'constant real numbers' and 'variable real numbers', have been clearly stated by Frege and need not be repeated here. The fact is that a satisfactory theory has never been developed on this basis, and it is not easy to see how it might be done.

At the time when he wrote this, Church was surely right.

Berkeley, Leśniewski, and others have formulated immensely influential general arguments against the very coherence of the concept *arbitrary object*. In Fine (1983, 1985b), Fine gave a much less influential but uncommonly incisive critique of these arguments. He showed us where the sceptical arguments go wrong. Frege asked pointed questions that a theory of arbitrary objects should answer. In this book, I have tried to provide principled answers to Frege's questions. More importantly, I have tried to take them to be of fundamental methodological importance for the development of arbitrary object theory. But Church's challenge to formulate a general, systematic, and fully articulated theory of arbitrary objects has not yet been met – not completely, anyway.

Fine identified key ingredients of a positive philosophical theory of arbitrary objects and *sketched* a theory of arbitrary objects. But his theory is not as systematic as would be desirable and is at places not developed in sufficient detail. In this monograph I have concentrated on laying the *groundwork* for general theory of arbitrary entities. I have incorporated what I believe to be correct in Fine's outline theory – i.e. most of it – and have departed from his view where I thought that it should be developed in a different direction. I have worked from examples – the natural numbers were my stock example – in order to be able to discern the contours of a theory of arbitrary objects, and to uncover constraints that such a theory should satisfy. I hope I have made some progress compared to what Fine has done earlier. But the result is still programmatic: the theory of arbitrary objects has not fully emerged yet.

In sum, I believe that the general metaphysical theory of arbitrary objects is still in an underdeveloped state, and that work on applications will only be fruitful if it is supported by fundamental (or 'pure') research in arbitrary object theory. For these reasons, I put the development of the general theory of arbitrary entities on the top of my list:

Problem 11.1. Articulate and defend a general metaphysical and logical theory of arbitrary entities.

11.3 Aspects and Applications

This section is concerned with problems of a more closely circumscribed nature. Some of the problems listed below are somewhat applied, some of them less so. I will concentrate on what may be fruitful problems for *my* theory of arbitrary objects as opposed to Fine's theory or Santambrogio's. Thus, in view of what was said in Section 7.7, the problem of formulating a general theory of Forms or universals in terms of arbitrary objects is not on the list.

The reader will have become painfully aware of the fact that this book contains many threads that were not pursued far and that could have been tied to other threads (that were also not pursued far). This gave rise to problems and open questions that I have tried to highlight at various places. Some of these problems are far from trivial. But most of them are smaller scale than the challenges than I will now touch upon.

11.3.1 History of mathematics.

Problem 11.2. Give a rational reconstruction of the nineteenth-century conception of mathematical variables.

The nineteenth-century conception of *variable numbers* is now in disrepute. Almost everyone regards it as obsolete, and many take it to be irredeemably inconsistent. In many respects, the present status of variable numbers is like that of infinitesimals before the discovery of non-standard analysis.

It is not clear to me that variable numbers deserve the bad press that they have received in the twentieth century. One way to change this would be to do for variable quantities what Robinson (1961) did for infinitesimals, i.e. to articulate a precise and coherent theory that makes sense of much of what nineteenth-century Analysts said informally about variable numbers and to make straightforward sense of most of the *mathematical* work that Analysts were doing with variable numbers. I have tried to take some first steps in this direction in Section 3.9 and in Section 4.1.3.

I have heard the project of describing Calculus in terms of variable quantities mentioned almost in the same breath as the project of giving a theory of infinitesimals in terms of variable quantities. But as I remarked in Section 6.4, the rational reconstruction of the concept of variable mathematical quantities should not *automatically* be tied to the concept of infinitesimals. They seem, at least at first blush, two different (albeit not unrelated) projects.

11.3.2 Degrees.

Problem 11.3. Develop a theory of degrees of arbitrariness.

Some first steps towards addressing this problem are taken in Santambrogio (1987). I myself have done little more than to suggest (in Section 9.4) some alternative concepts around which research pertaining to this question may be organised and to catalogue some of the basic properties of these concepts. There are hints that some of the concepts that I have identified may give rise to intricate and interesting degree structures that have not yet been investigated. I have in this context also touched upon possible connections with concepts of randomness that apply to numbers. But this theme, too, was not pursued far.

11.3.3 Dependence.

Problem 11.4. Model Fine's theory of arbitrary objects in the framework of dependence logic and its relatives.

Dependence logic (Väänänen 2007) and variations on it such as independence logic (as described in Väänänen and Grädel 2013) appear to be a suitable frameworks for modelling Fine's theory of arbitrary objects, which is organised around an irreducible concept of dependence. Some authors have noticed this (for instance San Gines 2014), but as far as I know, this connection has so far not been investigated in detail.

11.3.4 Forcing.

Problem 11.5. Explore the connections between arbitrary sets and forcing.

This is a potential avenue of future research that already features on Fine's list (Fine 1985b, pp. 45–46). I did not pursue this question because it seemed to me premature at the current stage in the development of arbitrary object theory. But it remains on the list. Especially the connection with the Boolean valued models approach (see Bell 2005) merits attention.

11.3.5 Carnapian proof systems.

Problem 11.6. Investigate proof-theoretic aspects of the formalisation of generic natural number systems in Carnapian modal logic.

We have seen in Chapter 9 that the 'smaller' generic natural number structures (i.e., the system of Section 4.1), or, from a proof theoretic point of view equivalently, the computable generic natural number structure (of Section 6.6) are from a proof theoretic perspective more tractable than the theory of the full generic natural number structure. And this is so despite the fact that they are equally expressive. The proof theoretic properties of the resulting proof system (in Carnapian modal logic) for the computable generic natural number structure are at present completely unknown, and merit our attention. A key component here is the principle of transfinite induction that was formulated at the end of Chapter 9 (axiom 9.49).

11.3.6 New probabilistic methods?

Problem 11.7. Can the treatment of arbitrary natural numbers as random variables be related to uses of probabilistic methods in discrete mathematics?

Since the early work of Erdős and others, probabilistic methods have played a central role in discrete mathematics: see, for instance, Alon and Spencer (2000). One might therefore wonder if non-Archimedean notions of probability can be fruitfully related to probabilistic proof methods in discrete mathematics. It is probably needless to say that this challenge is a bit of a long shot.

11.3.7 New mathematical axioms?

Problem 11.8. Find probabilistic principles concerning sets that have large cardinal strength and/or consistency strength.

Freiling made a notorious attempt to decide the Continuum Hypothesis on the basis of a *probabilistic* principle concerning sets of real numbers (Freiling 1986). Most set theorists remain unconvinced by his particular argument. But the strategy is interesting. In the light of Section 10.5, one might wonder if non-Archimedean probability principles can be found that have some intrinsic plausibility and that have large cardinal strength and/or consistency strength.

Bibliography

Antonutti Marfori, Marianna, *Naturalising mathematics: a critical look at the Quine-Maddy debate*, Disputatio **4** (2012), 323–342.

Anselm, St., *Proslogion*, in S. N. Deane, *St. Anselm: basic writings*. Trans. by Sidney D. Deane (1962), Open Court, 1977.

Alon, Noga and Joel Spencer, *The probabilistic method*. 2nd edn, Wiley, 2000.

Bacon, John, *The untenability of genera*, Logique et Analyse **65** (1974), 197–208.

Bell, John, *Set theory: Boolean-valued models and independence proofs*. Clarendon Press, 2005.

Bell, John and Alan Slomson, *Models and ultraproducts*. 2nd edn, Dover, 2006.

Belnap, Nuel and Thomas Müller, *CIFOL: case-intensional first-order logic. I: toward a theory of sorts*, Journal of Philosophical Logic **43** (2014), 393–437.

Benacerraf, Paul, *What numbers could not be*, Philosophical Review **74** (1965), 47–73.

    Mathematical truth, Journal of Philosophy **70** (1973), 661–679.

    What mathematical truth could not be – I, in A. Morton and S. Stich (eds), *Benacerraf and his critics*, pp. 9–59, Blackwell, 1996.

Benci, Vieri and Mauro Di Nasso, *Numerosities of labelled sets: a new way of counting*, Advances in Mathematics **17** (2003), 50–67.

Benci, Vieri, Marco Forti, and Mauro Di Nasso, *An Euclidean measure of size for mathematical universes*, Logique et Analyse **50** (2007), 43–62.

Benci, Vieri, Leon Horsten, and Sylvia Wenmackers, *Non-Archimedean probability*, Milan Journal of Mathematics **81** (2013), 121–151.

    Infinitesimal probabilities, British Journal for the Philosophy of Science **69** (2018), 509–552.

Berkeley, George, *A treatise concerning human knowledge*. Edited by D. R. Wilkins (2002), Dublin, 1710.

Blitzstein, Joseph and Jessica Hwang, *Introduction to probability*. CRC Press, 2015.

Boolos, George, *Nominalist platonism*, Philosophical Review **94** (1985), 327–344.

Boolos, George, Richard Jeffrey, and John Burgess, *Computability and logic*. 5th edn, Cambridge University Press, 2007.

Breckenridge, Wylie and Ofra Magidor, *Arbitrary reference*, Philosophical Studies **158** (2012), 377–400.

Brickhill, Hazel and Leon Horsten, *Triangulating non-Archimedean probability*, Review of Symbolic Logic **11** (2018), 519–546.

Sets and probability, arXiv:1903.08361, 2019, 26p.

Burgess, John and Gideon Rosen, *A subject with no object: strategies for nominalistic interpretation of mathematics.* Clarendon Press, 1997.

Bressan, Aldo, *A general interpreted modal calculus*, Yale University Press, 1972.

Burgess, John, *Identity, indiscernibility, and ante rem structuralism. Book Review: Stewart Shapiro, Philosophy of mathematics: structure and ontology*, Notre Dame Journal of Formal Logic **40** (1999), 283–291.

Mathematics and Bleak House, Philosophia Mathematica **12** (2004a), 18–36.

Quine, analyticity and philosophy of mathematics, Philosophical Quarterly **54** (2004b), 38–55.

Critical study / book review: Charles Parsons. Mathematical thought and its objects, Philosophia Mathematica **16** (2008), 402–420.

Philosophical logic. Princeton University Press, 2009.

Rigor and structure. Oxford University Press, 2015.

Parsons and the structuralist view, in O. Rechter (ed), *Intuition and reason*, Forthcoming.

Button, Tim and Sean Walsh, *Philosophy and model theory.* Oxford University Press, 2018.

Cameron, Peter, *The random graph*, in R. Graham and J. Nešetřil (eds), *The mathematics of Paul Erdős II. Algorithms and combinatorics, vol. 14*, pp. 333–351, Springer, 1997.

Cantor, Georg, *Mitteilungen zur Lehre vom Transfiniten*, Zeitschrift für Philosophie und Philosophische Kritik **91** (1887), 81–125.

Carnap, Rudolf, *Empiricism, semantics, and ontology*, Revue Internationale de Philosophie **4** (1950), 40–50.

Meaning and necessity: a study in semantics and modal logic. 3rd edn, University of Chicago Press, 1956.

Chaitin, Gregory, *A theory of program size complexity formally identical to information theory*, Journal of the Association of Computing Machinery **22** (1975), 329–340.

Church, Alonzo, *Introduction to mathematical logic.* Vol. 1. Princeton University Press, 1962.

Damjanovic, Zlatan, *Mutual interpretability of Robinson arithmetic and adjunctive set theory with extensionality*, Bulletin of Symbolic Logic **23** (2017), 381–404.

Dedekind, Richard, *Was sind und was sollen die Zahlen?* Vieweg, 1888.

de Finetti, Bruno, *Theory of probability.* 2 vols. Wiley, 1974.

de la Vallée Poussin, Charles-Jean, *Cours d'analyse infinitésimale.* Tome I, A. Uystpruyst-Dieudonné, 1903.

Diestel, Richard, *Graph theory.* Springer, 2006.

Dieterle, Jill, *Mathematical, astrological, and theological naturalism*, Philosophia Mathematica **7** (1999), 129–135.

Dummett, Michael, *The logical basis of metaphysics.* Harvard University Press, 1993.

Erdős, Paul and Alfréd Rényi, *Asymmetric graphs*, Acta Mathematica Academiae Scientiae Hungaricae **14** (1963), 295–315.

Euler, Leonhard, *Institutiones calculi differentialis. Vol. 1, Opera omnia, ser. 1, vol. 10 (1913)*, Acadamiae Imperialis Scientiarum, St. Petersburg, 1755.

Evans, Gareth, *Can there be vague objects?*, Analysis **38** (1978), 208.

Ferreira, Fernando and Gilda Ferreira, *Interpretability in Robinson's Q*, Bulletin of Symbolic Logic **19** (2013), 289–317.

Feyerabend, Paul, *Against method*. Verso, 1975.

Field, Hartry, *Science without numbers*. Blackwell, 1980.

    Is mathematical knowledge just logical knowledge?, Philosophical Review **93** (1984), 509–552.

Fine, Arthur, *And not anti-realism either*, Noûs **18** (1984a), 51–65.

    The natural ontological attitude, in J. Leplin (ed), *Scientific realism*, pp. 261–277, University of California Press, 1984b.

Fine, Kit, *Vagueness, truth and logic*, Synthese **30** (1975), 265–300.

    A defence of arbitrary objects. I: Kit Fine, Proceedings of the Aristotelian Society, Supplementary Volumes **57** (1983), 55–77.

    Natural deduction and arbitrary objects, Journal of Philosophical Logic **14** (1985a), 57–107.

    Reasoning about arbitrary objects. Basil Blackwell, 1985b.

    Cantorian abstraction: a reconstruction and defense, Journal of Philosophy **95** (1998), 599–634.

    The question of realism, Philosophical Imprints **1** (2001), 1–30.

    Class and membership, Journal of Philosophy, **102**, 547–572.

    Our knowledge of mathematical objects, in T. Szabo and J. Hawthorne (eds), *Oxford studies in epistemology*, vol. 1, pp. 89–110, Oxford University Press, 2005b.

    Form, Journal of Philosophy **114** (2017a), 509–535.

    Naive metaphysics, Philosophical Issues **1** (2017b), 98–113.

Frege, Gottlob, *The foundations of arithmetic: a logico-mathematical enquiry into the concept of number*. 2nd edn. Trans. by J.L. Austin (1968), Northwestern, 1984.

    Philosophical writings of Gottlob Frege. Trans. by P. Geach and M. Black, Blackwell, 1960.

    Posthumous writings. Trans. by P. Long and R.M. White, Blackwell, 1979.

Freiling, Chris, *Axioms of symmetry: throwing darts at the real number line*, Journal of Symbolic Logic **51** (1986), 190–200.

Garson, James, *Modal logic for philosophers*. Cambridge University Press, 2006.

Gödel, Kurt, *What is Cantor's continuum problem?*, American Mathematical Monthly **54** (1947), 515–525.

Goldblatt, Robert, *Lectures on the hyperreals: an introduction to non-standard analysis*. Springer, 1998.

Halbach, Volker and Leon Horsten, *Computational structuralism*, Philosophia Mathematica **13** (2006), 174–186.

Hamkins, Joel, *The set theoretic multiverse*, Bulletin of Symbolic Logic **5** (2012), 416–449.

Hazen, Alan, *Review of Crispin Wright: Frege's conception of numbers as objects*, Journal of Philosophy **63** (1985), 250–254.

Heidegger, Martin, *Introduction to 'What is metaphysics' ('Getting to the bottom of metaphysics'). Trans. by M. Groth*, 1949.

Hellman, Geoffrey, *Mathematics without numbers: towards a modal-structural interpretation*. Clarendon Press, 1993.

  Structuralism, in S. Shapiro (ed), *Oxford handbook of philosophy of mathematics and logic*, pp. 536–562, Oxford University Press, 2006.

Heylen, Jan, *Carnap's theory of descriptions and its problems*, Studia Logica **94** (2010), 355–380.

Hilbert, David, *Naturerkennen und Logik*, in D. Hilbert, *Gesammelte Abhandlungen. Dritter Band (1935)*, pp. 378–387, Julius Springer, 1930.

  Neubegrundung der Mathematik, in D. Hilbert, *Gesammelte Abhandlungen. Dritter Band*, pp. 157–177, Julius Springer, 1935.

Horsten, Leon, *Canonical naming systems*, Minds and Machines **15** (2005), 229–257.

  Levity, Mind **118** (2009), 555–581.

  Vom Zählen zu den Zahlen. On the relation between computation and arithmetical structuralism, Philosophia Mathematica **20** (2012), 275–288.

  Mathematical philosophy?, in H. Andersen et al. (eds), *The philosophy of science in a European perspective*, vol. 4, pp. 73–86, Springer, 2013.

  Generic structuralism, Philosophia Mathematica, online first (2019), 19p.

Horsten, Leon and Stanislav Speranski, *Reasoning about arbitrary natural numbers from a Carnapian perspective*, Journal of Philosophical Logic (forthcoming).

Hume, David, *A treatise of human nature. Being an attempt to introduce the experimental method of reasoning into moral subjects (1986)*. Penguin, 1739.

Incurvati, Luca, *On the concept of finitism*, Synthese **129** (2015), 2413–2436.

Isaacson, Daniel, *The reality of mathematics and the case of set theory*, in S. Novak and A. Simonyi (eds), *Truth, reference, and realism*, pp. 1–76, Central European University Press, 2011.

Ito, Ryo, *Russell's metaphysical accounts of logic*. PhD dissertation, University of St Andrews, 2017.

Jeshion, Robin, *Intuiting the infinite*, Philosophical Studies **171** (2014), 327–349.

Kaye, Richard, *Models of Peano Arithmetic*. Clarendon Press, 1991.

Kearns, Stephen and Ofra Magidor, *Epistemicism about vagueness and metalinguistic safety*, Philosophical Perspectives **22** (2008), 277–304.

Keränen, Jukka, *The identity problem for realist structuralism*, Philosophia Mathematica **9** (2001), 308–330.

Ketland, Geoffrey, *Structuralism and the identity of indiscernibles*, Analysis **66** (2006), 303–315.

King, Jeffrey, *Instantial terms, anaphora and arbitrary objects*, Philosophical Studies **61** (1991), 239–265.

Kremer, Philip, *Indeterminacy of fair infinite lotteries*, Synthese **191** (2014), 1757–1760.

Kripke, Saul, *Semantical considerations on modal logic*, Acta Philosophica Fennica **16** (1963), 83–94.

 Naming and necessity. Harvard University Press, 1980.

 Individual concepts: their logic, philosophy, and some of their uses, Proceedings and Addresses of the American Philosophical Association **66** (1992), 70–73.

Kuhn, Thomas, *Objectivity, value judgement and theory choice*, in Th. Kuhn, *The essential tension*, pp. 320–339, University of Chicago Press, 1977.

Ladyman, James and Don Ross, *Everything must go: metaphysics naturalized*. Oxford University Press, 2009.

Ladyman, James, Øystein Linnebo, and Richard Pettigrew, *Identity and discernibility in philosophy and logic*, Review of Symbolic Logic **5** (2012), 162–186.

Lakatos, Imre, *Falsfication and the methodology of scientific research programmes*, in I. Lakatos and A. Musgrave (eds), *Cricism and the growth of knowledge*, pp. 91–196, Cambridge University Press, 1970.

Lawvere, William, *Functorial semantics for algebraic theories*. PhD dissertation, Columbia University, 1963.

Leitgeb, Hannes and James Ladyman, *Criteria of identity and structuralist ontology*, Philosophia Mathematica **16** (2008), 388–396.

Lewis, David, *On the plurality of worlds*. Blackwell, 1986.

 Parts of classes. Basil Blackwell, 1991.

Linnebo, Øystein, *Philosophy of mathematics*. Princeton University Press, 2017.

Littlewood, J., *Littlewood's miscellany*. Edited by B. Bollobas, Cambridge University Press, 1986.

 Thin objects: an abstractionist account. Oxford University Press, 2018.

Linnebo, Øystein and Richard Pettigrew, *Two types of abstraction for structuralism*, Philosophical Quarterly **64** (2014), 267–283.

Locke, John, *An essay concerning human understanding*. Edited by A. C. Fraser (1894), Clarendon Press, 1690.

Loux, Michael and Dean Zimmerman (eds), *The Oxford handbook of metaphysics*. Oxford University Press, 2003.

MacBride, Fraser, *Structuralism reconsidered*, in S. Shapiro (ed), *Oxford handbook of philosophy of mathematics and logic*, pp. 563–589, Oxford University Press, 2005.

Macnamara, John, *Review: Kit Fine. Reasoning with arbitrary objects*, Journal of Symbolic Logic **53** (1988), 305–306.

Maddy, Penelope, *Naturalism in mathematics*. Oxford University Press, 1997.

Second philosophy: a naturalistic method. Oxford University Press, 2007.

Manders, Kenneth, *Domain extensions and the philosophy of mathematics*, Journal of Philosophy **86** (1989), 553–562.

Martens, David, *Combination, convention, and possibility*, Journal of Philosophy **103** (2006), 577–586.

Martin, Donald, *Multiple universes of sets and indeterminate truth values*, Topoi **20** (2001), 5–16.

Mates, Benson, *Identity and predication in Plato*, Phronesis **24** (1979), 211–229.

Mayberry, John, *What is required for a foundation of mathematics?*, Philosophia Mathematica **2** (1994), 16–35.

McMullin, Ernan, *A case for scientific realism*, in J. Leplin (ed), *Scientific realism*, pp. 8–40, University of California Press, 1984.

Menger, Karl, *On variables in mathematics and in natural science*, British Journal for the Philosophy of Science **18** (1954), 134–142.

Meyer Viol, Wilfried, *Instantial logic: an investigation into reasoning with instances.* ILLC, 1995.

Myhill, John, *Review: W. V. Quine, On Carnap's views on ontology*, Journal of Symbolic Logic **20** (1955), 61–62.

Nelson, Edward, *Predicative arithmetic.* Princeton University Press, 1986.

Nickel, Benhard, *Generics*, in B. Hale, A. Miller, and C. Wright (eds), *The Blackwell companion to the philosophy of language*, 2nd edn, pp. 437–462, Blackwell, 2017.

Niebergall, Karl-Georg, *On the logic of reducibility: axioms and examples*, Erkenntnis **53** (2000), 27–62.

Nodelman, Uri and Edward Zalta, *Foundations for mathematical structuralism*, Mind **123** (2014), 39–78.

Oliver, Alex and Oliver Smiley, *What are sets and what are they for?*, Philosophical Perspectives **20** (2006), 123–155.

Parsons, Charles, *Mathematical intuition*, Proceedings of the Aristotelian Society, Supplementary Volumes **80** (1980), 145–168.

The structuralist view of mathematical objects, Synthese **84** (1990), 303–346.

Structuralism and metaphysics, Philosophical Quarterly **54** (2004), 56–77.

Mathematical thought and its objects. Cambridge University Press, 2008.

Pettigrew, Richard, *Platonism and Aristotelianism in mathematics*, Philosophia Mathematica **16** (2008), 310–332.

Poincaré, Henri, *Science and hypothesis.* Dover, 1905.

Putnam, Hilary, *Mathematics without foundations*, Journal of Philosophy **67** (1967), 5–22.

Quine, William V.O., *On what there is*, Review of Metaphysics **2** (1948), 21–48.

Two dogmas of empiricism, Philosophical Review **60** (1951), 20–43.

Epistemology naturalized, in W.V.O. Quine, *Ontological relativity and other essays*, pp. 69–90, Columbia University Press, 1969.

The variable, in R. Parikh (ed), *Logic Colloquium 1972–1973*, pp. 155–168, Springer, 1975.

Theories and things. Harvard University Press, 1981.

Review of Parsons's 'Mathematics in philosophy', Journal of Philosophy **81** (1984), 783–794.

Philosophy of logic. 2nd edn, Harvard University Press, 1986.

Rescher, Nicholas, *Can there be random individuals?* Analysis **18** (1958), 114–117.

Resnik, Michael, *Mathematics as a science of patterns: ontology and reference*, Noûs **15** (1981), 529–550.

Second order logic still wild, Journal of Philosophy **85** (1988), 75–87.

Robinson, Abraham, *Non-standard analysis*, Proceedings of the Royal Academy of Sciences, Amsterdam, Series A **64** (1961), 432–440.

Rosen, Gideon, *Metaphysical dependence: grounding and reduction*, in B. Hale and A. Hoffmann (eds), *Modality: metaphysics, logic, and epistemology*, pp. 109–136, Oxford University Press, 2010.

Rovelli, Carlo, *'Space is blue and birds fly through it'*, Philosophical Transactions of the Royal Society A **376** (2018), no. 2017.0312, 12p.

Russell, Bertrand, *On meaning and denotation*, in A. Urquhart (ed), *The collected papers of Bertrand Russell 4: foundations of logic 1903–1905*, pp. 314–358, Routledge, 1903a.

The principles of mathematics (1996). W. W. Norton, 1903b.

On denoting, Mind **56** (1905), 479–493.

Knowledge by acquaintance and knowledge by description, Proceedings of the Aristotelian Society **11** (1910), 108–128.

Introduction to mathematical philosophy. George Allen & Unwin, 1919.

San Gines, Aranzanzu, *On Skolem functions, and arbitrary objects: an analysis of Kit Fine's mysterious claim*, Teorema **33** (2014), 137–150.

Santambrogio, Marco, *Generic and intensional objects*, Synthese **73** (1987), 637–663.

Review: Reasoning about arbitrary objects, by K. Fine, Noûs **22** (1988), 630–635.

Was Frege right about variable objects?, in K. Mulligan (ed), *Language, truth and ontology*, pp. 133–156, Kluwer, 1992.

Schiemer, Georg and John Wigglesworth, *The structuralist thesis reconsidered*, British Journal for the Philosophy of Science, forthcoming.

Schiffer, Jonathan, *The things we mean*. Clarendon Press, 2003.

Schoenfield, Joseph, *The axioms of set theory*, in J. Barwise (ed), *Handbook of mathematical logic*, pp. 321–345, North-Holland, 1977.

Shafer, Glenn, *When to call a variable random*, working paper no 41 of the Game-Theoretic Probability and Finance Project (2018), 64p.

Shapiro, Stewart, *Acceptable notation*, Notre Dame Journal of Formal Logic **23** (1982), 14–20.

Foundations without foundationalism: a case for second-order logic. Oxford University Press, 1991.

Philosophy of mathematics: structure and ontology. Oxford University Press, 1997.

Structure and identity, in F. MacBride (ed), *Identity and modality,* pp. 34–69, Oxford University Press, 2006.

Identity, indiscernibility, and ante rem structuralism: the tale of i and -i, Philosophia Mathematica **16** (2008), 285–309.

Shapiro, Stuart C., *A logic of arbitrary and indefinite objects,* in D. Dubois et al. (eds), *Principles of knowledge representation and reasoning: proceedings of the ninth international conference (KR2004),* pp. 265–275, AAAI Press, 2004.

Sider, Theodore, *Four-dimensionalism: an ontology of persistence and time.* Oxford University Press, 2001.

Writing the book of the world. Oxford University Press, 2011.

Sklar, Lawrence, *Space, time, and spacetime.* University of California Press, 1975.

Skolem, Thoralf, *Some remarks on axiomatized set theory,* in J. van Heijenoort (ed), *From Frege to Gödel: a source book in mathematical logic, 1879–1931* (1967), pp. 252–263, Harvard University Press, 1922.

Solomonoff, Ray, *A preliminary report on a general theory of inductive inference,* Report v-131, Zator Co., 1960.

Strawson, Peter, *Individuals.* Methuen, 1959.

Entity and identity, in P. Strawson, *Entity and identity and other essays (1997),* pp. 21–51, Clarendon Press, 1976.

Švejdar, Vítězslav, *On interpretability in the theory of concatenation,* Notre Dame Journal of Formal Logic **50** (2009), 87–95.

Szubka, Tadeusz, *Metaontological maximalism and minimalism: Fine versus Horwich,* in A. Kuzniar and J. Odrowaz-Sypniewska (eds), *Uncovering facts and values: studies in contemporary epistemology and political philosophy.* Poznan Studies in the Philosophy of Science and the Humanities 107, 2016.

Tait, William, *Finitism,* Journal of Philosophy **78** (1981), 524–546.

Truth and proof: the platonism of mathematics, in W. Tait, *The provenance of pure reason: essays in the philosophy of mathematics and its history (2005),* pp. 61–88, Oxford University Press, 1986.

Beyond the axioms: the question of objectivity in mathematics, Philosophia Mathematica **9** (2001), 21–36.

Tennant, Neil, *A defence of arbitrary objects. II: Neil Tennant,* Proceedings of the Aristotelian Society, Supplementary Volumes **57** (1983), 79–89.

Troelstra, Anne, *Choice sequences: a chapter in intuitionistic mathematics.* Clarendon Press, 1977.

Truss, John, *Foundations of mathematical analysis.* Oxford University Press, 1997.

Väänänen, Jouko, *Dependence logic: A new approach to independence-friendly logic.* Cambridge University Press, 2007.

Väänänen, Jouko and Erich Grädel, *Dependence and independence*, Studia Logia **101** (2013), 399–410.

van Fraassen, Bas, *The scientific image*. Oxford University Press, 1980.

Visser, Albert, *Growing commas: a study of sequentiality and concatenation*, Notre Dame Journal of Formal Logic **50** (2009), 61–85.

Williamson, Timothy, *Modal logic as metaphysics*. Oxford University Press, 2013.

Wittgenstein, Ludwig, *Philosophical investigations*. Blackwell, 1953.

Worrall, John, *Structural realism: the best of both worlds?*, Dialectica **43** (1989), 99–124.

Zermelo, Ernst, *On boundary numbers and domains of sets* (translated by W. Hallett), in W. Ewald (ed), *From Kant to Hilbert: a source book in mathematics*. Vol. 2 (1996), pp. 1208–1233, Oxford, 1930.

Index

Page numbers in boldface refer to key occurrences of words

Printed in the United States
by Baker & Taylor Publisher Services